Security Relationship Management

Aligning information security to the goals and strategies of the business is paramount for ensuring risks are addressed, without an abundance of negative impacts to the company. But how does a Chief Information Security Officer (CISO) accomplish effective alignment? A security executive must understand the detailed needs of business leaders and stakeholders from across all corners of the company. We cannot rely on a standard cadence of general security discussions across all of the lines of business, as well as functional areas, and expect our alignment to be maximally effective. Instead, we should promote our security programs in such a way that makes it personal to whomever we are speaking with at any given time.

By leveraging already established and tested marketing concepts, slightly altered for information security, the CISO can tailor their message to fit the needs of each stakeholder. This allows for in-depth business alignment, as well as a holistic view of the company's underpinnings for the CISO. Within these pages, the reader will learn how segmentation, the Four P's, and customer relationship management techniques can help to transform their security program. Additionally, the book introduces a concept called Security Relationship Management (SRM) that optimizes the creation and nurturing of the hundreds of professional relationships (within and outside the company) that a CISO must balance each week. Through structured tracking of interactions and analyzing SRM data, the CISO ensures that relationships are managed effectively, which increases alignment between the business and cybersecurity initiatives. Pick up your copy of *Security Relationship Management: Leveraging Marketing Concepts to Advance a Cybersecurity Program* today to begin your SRM journey.

Please visit www.novelsecurity.com for more information.

Security, Audit and Leadership Series

Series Editor: Dan Swanson, Dan Swanson and Associates, Ltd.,
Winnipeg, Manitoba, Canada.

The ***Security, Audit and Leadership Series*** publishes leading-edge books on critical subjects facing security and audit executives as well as business leaders. Key topics addressed include Leadership, Cybersecurity, Security Leadership, Privacy, Strategic Risk Management, Auditing IT, Audit Management and Leadership

Continuous Auditing with AI in the Public Sector
Lourens Erasmus and Sezer Bozkus Kahyaoglu

Ironwill 360° Leadership: Moving Forward: Unlock Twelve Emerging Trends for Forward Thinking Leaders
Douglas P. Pflug

The CISO Playbook
Andres Andreu

Leveraging Blockchain Technology: Governance, Risk, Compliance, Security, and Benevolent Use Cases
Shaun Aghili

The Closing of the Auditor's Mind?: How to Reverse the Erosion of Trust, Virtue, and Wisdom in Internal Auditing
David J. O'Regan

Radical Reporting: Writing Better Audit, Risk, Compliance, and Information Security Reports (Second Edition)
Sara I. James

Team Intelligence: A New Method Using Swarm Intelligence for Building Successful Teams
Mohammad Nozari

The Gardener of Governance: A Call to Action for Effective Internal Auditing
Rainer Lenz and Barrie Enslin

Navigating the Cyber Maze: Insights and Humor on the Digital Frontier
Matthias Muhlert

Safeguarding the Future: Security and Privacy by Design for AI, Metaverse, Blockchain, and Beyond
Alan Tang

Security Relationship Management: Leveraging Marketing Concepts to Advance a Cybersecurity Program
Lee Parrish

Security Relationship Management

Leveraging Marketing Concepts to Advance a Cybersecurity Program

Lee Parrish

CRC Press
Taylor & Francis Group
Boca Raton London New York

CRC Press is an imprint of the
Taylor & Francis Group, an **informa** business

Designed cover image: Shutterstock

First edition published 2025
by CRC Press
2385 NW Executive Center Drive, Suite 320, Boca Raton FL 33431

and by CRC Press
4 Park Square, Milton Park, Abingdon, Oxon, OX14 4RN

CRC Press is an imprint of Taylor & Francis Group, LLC

ISBN: 978-1-032-97998-4 (hbk)
ISBN: 978-1-032-98184-0 (pbk)
ISBN: 978-1-003-59748-3 (ebk)

DOI: 10.1201/9781003597483

Typeset in Caslon
by SPi Technologies India Pvt Ltd (Straive)

For Nancy, Tania, and Grady: Thank you for your support and the sacrifices you've made (and continue to make) on my journey.

Always and forever…

Contents

Author

Lee Parrish is an award-winning technology executive with over two decades of unique experience in blending cybersecurity expertise with essential business competencies. As a Chief Information Security Officer, he has built customized cybersecurity strategies for global Fortune 500 corporations and has led real-world incident responses to cyber events. Lee has served as a trusted advisor on cybersecurity to multiple boards consisting of chief executive officers, a former White House Chief of Staff, retired high-ranking military officers, and a former U.S. Presidential candidate.

Lee possesses two graduate degrees and is certified as both a Boardroom Qualified Technology Expert and a Certified Information Systems Security Professional. He has published numerous articles in industry journals, contributed to a best-selling information security book, and authored *The Shortest Hour: An Applied Approach to Boardroom Governance of Cyber Security*, as well as a children's book on cybersecurity. He is a frequent speaker at international security conferences and a guest on various podcasts.

Lee is a combat veteran of the United States Marine Corps.

Introduction

Arkansas was still reeling from one of the largest ice storms to ever hit Little Rock as I was settling into my first role as an information security executive. The ice that had accumulated on every tree was beginning to thaw and power was almost fully restored to the thousands of residents who were in the dark for weeks. All against a backdrop of a pre-9/11 United States.

I began to reach out to cybersecurity peers in the local community to garner tips that could assist me in building my own program and was very excited to speak to the Chief Information Security Officer (CISO) for a big corporation occupying one of the larger office buildings in downtown Little Rock. As I would find out in the next twenty-four years of my career, members of our cybersecurity industry are always willing to help one another, and he was kind enough to put together a thoughtful day of tours and introductions through the various departments of his company.

As we walked, he stopped and introduced me to the Vice President of Physical (Corporate) Security for the company. After the three of us casually chatted about my new role and the importance of cyber security, my host told the security executive,

"Actually, I don't believe *we've* ever met in person ourselves", and then formally introduced himself to the Vice President.

We then ran into the Chief Financial Officer (CFO) and the same thing happened; he had not formally met the CFO face to face until he introduced me. This scenario continued to play out, this time with human resources; all from a CISO who had been in his role for over a year. Sensing my surprise, he explained, "Most of my interactions with these executives are done through email, but much of the time I work closely with the directors on their teams".

This was a defining moment for me all those years ago insofar as I could not fathom how a security executive could be effective in their role without having a close, solid relationship with the key stakeholders of their company. I asked him about his background, and like many other CISOs at that time, he told me he was tapped for the role after many years in a successful technical career, hands on keyboard. It was then that the first thoughts about applying business concepts to information security came to mind, as well as the role of relationship building, in creating and maintaining a security program.

A few weeks later, I enrolled in the Executive MBA program at the University of Arkansas (Little Rock) and began my work in that curriculum. Peers and friends asked me at the time if my academic decision meant I was getting out of security and I told them I wasn't; my goal was to learn more about business so that I could do security better. The curriculum did not disappoint. It wasn't a year into the program, and I found myself traveling to Budapest, Vienna, Prague, and Guadalajara to tour various businesses, observe their operations and financial models, and understand how all aspects of their businesses efficiently work together to deliver products and services to their customers. Watching and talking with employees at the different companies as they made cosmetics, tequila, cardboard cartons, and other products, this MBA program placed me directly on the manufacturing floor, where my real education took place.

I was more energized than ever, and I applied what I learned to my CISO role back at home. I began reaching out to the Chief Executive Officer (CEO), Chief Human Resources Officer (CHRO), and CFO to understand their priorities. The CFO invited me to be a member of his risk committee made up of representatives from accounting, physical security, IT, and business operations. In my first meeting with them, he asked me what I would like to gain from my involvement on the committee. I expect he was anticipating an answer that inquired about the systems that drove the most critical business processes or perhaps a question on patching. He, as well as the other members of the committee, was surprised when I said that I would like to understand the budgets for each of the functional areas that made up the risk committee – costs that would include total labor cost, travel/entertainment, outsourcing, capital, expense, and special projects.

A smile slowly grew on his face (I was speaking his language) when he realized where I was going with my request. The following month, the members of the committee brought back the financial data and the results were enlightening. We overlaid the spend with the top enterprise risks that were defined in previous meetings and concluded that we needed to reallocate resources across our different functional areas to address these risks more effectively. I was quickly realizing that to be successful, I needed to continue down the path of growing relationships and applying business concepts to my cybersecurity program.

I spoke with people at all levels of my company to understand every facet of the corporation. With each conversation, there was at least some correlation between cybersecurity and their area of focus. Things were going fantastic, and I realized this is what I wanted to do for the rest of my career. As my relationship work continued, it was the fourth year into my inaugural security executive role that I was recruited for a CISO position with an operating sector of a Fortune 500 corporation in the aerospace/defense industry. I flew to Washington DC on a sunny Thursday morning to interview with the hiring manager for the information security program (he would later become a valued colleague throughout my career). We talked about the role of the sector CISO, and I told him about my accomplishments of incorporating business into the cybersecurity program and he agreed, saying this sector CISO role was akin to an inside sales position. Boarding my flight home to Little Rock, it was apparent to me that this company, and this leader, understood security. I took the position, moved to Gulfport, Mississippi, and it was one of the most fulfilling roles of my career.

After a year with my new company, I had a desire to do more in the security industry. I wanted to publish an article in a security trade journal on this notion of deeply integrating business into security, but I did not have the contacts to assist me in being published so early in my career. I returned one weekend to the Arkansas apartment that my family and I retained in our previous hometown. I spent the entire time writing an article that outlined the need for integrating business concepts into cybersecurity. Once I completed the article, I sent it (unsolicited) to the publisher of a popular information security magazine, then loaded up my car for the long drive back to the Mississippi Gulf Coast. As I pulled into my driveway at home seven hours later, I received an email on my Blackberry that the publisher wanted to publish my article in an upcoming issue. Publishing that small article over twenty years ago helped to validate that my opinions wanted to be heard in a professional forum, even if it was just a small step.

The turning point in my thoughts on security relationship management came in 2011 at a major cybersecurity conference in London where I gave a session on leveraging marketing concepts to advance a cybersecurity program. The session received some positive local press in the U.K., and the attendees provided me with very kind feedback, as well as ideas that I had not thought of during my preparation for the session. I left the conference feeling energized and boarded my flight from

London Heathrow with a million thoughts racing through my mind. During the flight, I scribbled some notes that would later become the blueprint for this book, which is now being realized some thirteen years later.

A key component of my marketing presentation in London was about formally building relationships across all corners of the business (and externally with partners) to deliver security services, much in the same way companies build and maintain relationships with their customers. But instead of calling it Customer Relationship Management (CRM), I refer to it as *Security Relationship Management* (SRM).

During the autumn of 2015, I decided to formalize my ideas on SRM and deliver them through a cloud-based enterprise application. I created a C-corporation, self-funded the start-up, and began developing the company on nights and weekends. A year later, I had a minimally viable product that would showcase the ability for CISOs to track internal and external relationships, and then examine the analytics related to their relationships. Peers in the security industry were kind enough to introduce me to venture capitalist firms in San Francisco and Menlo Park where I pitched the concept of SRM, and the platform. The feedback from the potential investors was very helpful, and I made several revisions based on their observations. I purchased booth space at a large security conference in San Francisco to demonstrate the product, and while there was good interest, I was competing with (at the time) sweeping industry trends addressing things like threat intelligence, user behavioral analysis, and code analysis (to name a few). I ultimately disbanded the start-up, realizing that our industry wasn't quite prepared for the concept of SRM nearly ten years ago.

Today, I am taking the lessons I have learned from over two decades as a CISO, along with experiences from my start-up, to promote the discipline of SRM through the pages of this book. The concepts I'll describe in the following chapters have worked well for me in my career, providing me not only with increased effectiveness in delivering security services, but a holistic view of the company through a more vivid lens. My hope is that this book provides you with some useful tips that you can use in your own program, and perhaps in some small way, inspire industry change on how we approach cybersecurity. Or at the very least, we'll secure some great relationships along the way.

Lee Parrish
October 2024

For more information, please visit: www.novelsecurity.com

1

Aligning on the CISO Role

Before diving into how security relationship management (SRM) can help us chart a path to success for the Chief Information Security Officer (CISO) role, we must first agree on the role's destination – how we see it contributing to an organization in the future – beyond what we do now. I won't get into the history and evolution of the CISO role; we are all very familiar with this from other books, articles, and our own knowledge of our industry. Instead, let's imagine a long-term vision for the future of the CISO role and align on that.

I frequently observe other senior executives to identify and pull out various qualities from each of them that I believe are well worth emulating. Each has their own area of expertise, and their title reflects that specialty, i.e., Chief Marketing Officer, Chief Legal Officer, Chief Financial Officer, and so on. While there are inherent interdependencies between these executive roles, I have found that those leaders who regularly stretch outside of their domain to help set the strategic direction for the company are the most effective. These are the executives I watch closely to learn from and imagine how I might apply the same behaviors to my role as CISO.

For example, one of the CFOs I have worked with is very good at outlining the company's performance in financial matters such as improving our gross profit margins, or perhaps maintaining a strong cash flow – he is an expert in his field. But when I talk with him, he is well versed in other areas such as inventory, supply chain, and manufacturing (to name a few) – not only how they impact corporate finance, but an overall knowledge of their

DOI: 10.1201/9781003597483-1

importance to enterprise success. He is part of an executive leadership team and provides insights into corporate strategies and makes enterprise-wide decisions, even outside his area of responsibility. He is a member of the executive leadership team and an active participant in the strategic direction for the company.

I believe that we have only scratched the surface with regard to the contributions that the role of CISO can have for its company. Of course, we will continue to get better as an industry in aligning our cyber work to enable the business; we've already seen significant advancements in our ability to accomplish this over the past ten years. But while some industry pundits call for the demise of the CISO role, I see major opportunities. Consider the attributes of our role that do nothing but position us for an increased span of control within our organization.

The Importance of Information Security

I recognize that all areas of the business are important to the success of a corporation. However, there is something about cybersecurity that elevates our discipline to that of a critical success factor. It would not surprise me if the readers of this book have seen firsthand that information security falls at either number one or number two on their own company's enterprise risk management program. It is a systemic risk that ebbs and flows as a result of both internal and external factors. When a breach occurs, the business must slow (or halt) to address the incident. Business resiliency is a driving factor in making cyber a serious topic that must be addressed if the company is to be successful. This importance lends itself to making cyber just as important as our peer business disciplines.

Enterprise Reach

There is a significant amount of debate about where the role of CISO should report. Rather than add my thoughts to the debate here, I'll only offer that no matter which organization that a

CISO belongs to, they consistently support projects and initiatives outside of their home department.

For many CISOs, they report to the Chief Information Officer (CIO) and have a front-row seat to the information technology (IT) projects being delivered through that organization. However, some companies may have separate organizations outside of IT dedicated to digital technology and eCommerce. In some cases, I've seen a company leverage both a CIO and a Chief Technology Officer (CTO) in separate organizations to deliver technology services.

As a CISO, your responsibilities don't end at the IT organization's boundary; your team provides security services to all corners of the business that fall outside of IT – and not every function of the business has this unique visibility. Moreover, your team contributes to supporting all regions, perhaps internationally, to enable those areas. The international impact of the CISO role creates an opportunity for your contributions to have significant influence to the corporation.

Executive Level Access

With the emergence of more network computing, the topic of cybersecurity (and the criticality of it) has extended outside of technology teams and into the executive suite, as well as the boardroom. Executives realize the importance of information security in its business resiliency plans and as a result, want to understand what security teams are doing to identify and manage cyber risks. Equally interested are boards of directors (or board committees) who need to provide oversight of management's strategies to address these risks.

Security leaders, largely absent from boardrooms and executive conference rooms in the past, are now finding themselves shoulder to shoulder with the highest levels of corporate leadership to discuss this critical topic. CISOs have become a fixture at management meetings, which has given them a unique opportunity to contribute to the ongoing operations of the business.

Putting It All Together

The combination of the importance of information security, enterprise reach, and executive level access provides the CISO with tangible opportunities to take the role to heights not seen previously. The ability for the CISO to control the three attributes listed earlier is not within our power. But because they all have been fatefully aligned by external forces to bring importance to the CISO role, the only factor remaining to exploit this growth opportunity is our ability to capitalize on the moment through our own professional capabilities and willingness to advance the role.

In order to catapult our ability to seize this important opportunity, we need to increase our own self development in the areas of business acumen, as well as building effective relationships across the enterprise. By controlling these internal factors, those in which we do indeed have the power to address, we can now position the role as one of a peer executive who contributes to the strategic direction of our company. The alternative is for us to stay focused on the deeply technical underpinnings which spawned our role and rely on an executive proxy to translate the technical terms into business on our behalf. This alternate path will most certainly suppress our ability to elevate our role.

The first step in heightening our position comes from a mantra I learned from a mentor of mine back in the late 1990s. He told me, "If you aspire to be in a role and wish to be in a position of contributing more, act like you are already in that role". One of the remaining things left to do then in order to launch our CISO role to the aforementioned heights is to begin contributing as if we are already in that position. We must double-down on our ability to contribute more to the strategic direction of our companies, even outside of the realm of information security.

The reasons outlined here are why I believe SRM is so important. In many ways, I believe it is the key to unlock our success in elevating the position to where it can contribute even more. We'll enjoy a more holistic view of our company, understand how they all blend to drive competitive advantage, and acknowledge where opportunities lie. SRM will provide us with all of these things.

Let's dive into the concept now.

2
Security Relationship Management Defined

The ultimate goal for any company that sells products and services is (or should be) to get as close to the customer as possible. While some companies send out customer surveys or rely on sterile market analysis data provided at a cost, a successful brand will spend time getting to know their customers using strategies that place them alongside their customers, in direct contact with them. In doing so, they intimately understand how customers feel about their products, the opportunities to enhance them, and maybe generate ideas for new product features.

There are many innovative examples of companies using communities, clubs, owner groups, customer roadshows, or other unique methods, to obtain that in-person engagement with their customers. In the cybersecurity field (really, any field), companies may leverage an annual user conference to bring together their customers to discuss all things "enter product name here". I attend many of these conferences each year and see first-hand the sense of community they provide, all wrapped around the product or service delivered to the customer. At some of these events, the attendees can be very enthusiastic supporters, some forming subgroups and user clubs. At the same time, employees of the company are sitting side by side with these customers. The valuable insights that are garnered from these close interactions cannot be understated. This is as close to the customer that a company can get. The feedback is actionable for the company; future product enhancements can be a direct result of real customer insights, not from broad projections or generic studies.

DOI: 10.1201/9781003597483-2

Other forms of customer engagement exist in the form of online forums, social media, social media influencers, and other channels delivered through electronic means. In today's digital world, these vectors provide value in reaching out to customers and should be significantly leveraged by companies, but the investment of time in meeting with customers face to face will pay even more extraordinary dividends in the long-term marketing strategy. At its core, these engagement methods are just a few examples of customer relationship management (CRM).

CRM is a discipline made up of practices and technologies that companies use to manage and grow customer relationships. There are many different aspects to CRM, and by my own admission, I am certainly no expert on the topic. There are many wonderful books, online resources, and professionals who can do the topic more justice than I, and it is worth investing the time to learn more on the topic from those resources.

For our purposes here (and what can be debated by the CRM industry), I believe there may be at least two aspects to CRM: one of the components is more transactional in nature; i.e., my wife orders an art kit for our granddaughters and when it arrives, it was not the product she ordered. She returns it so that the company can rectify the mistake, which, in turn, renews customer satisfaction and may increase the likelihood of future, repeat purchases. While this area is critically important, especially for the retention of a company's consumer base, my focus for this book is more directed to the proactive aspects of CRM: understanding and building long-term relationships with customers to create brand loyalty and drive improvements to products and services.

We all have companies in mind when thinking about brand loyalty. As you read this, I am sure that more than a few companies come to your mind related to your own product choices. I have very demanding needs when it comes to the quality of a few certain products and services I use, and as a result, there are a handful of companies that I use exclusively – I am extremely loyal to their brands. To cement the importance of brand loyalty, try this simple exercise. Ask a couple of your colleagues or friends about their brand preferences: Coke or Pepsi? Chevy trucks or Ford trucks? The answers they provide are not of great importance to our exercise here, but it is the passion in which they answer the questions that one should observe. Brands matter.

In researching for this book, specifically CRM, I read definitions from more than a few CRM technology corporations that defined CRM as a *technology* or a *tool* to manage customer relationships. As with cyber security, technology is a critical component to be sure – but only a partial means to a larger end. CRM is much more than technology.

Consider Levi Strauss, a Bavarian-born merchant who arrived in San Francisco in 1853 to open a wholesale dry goods business. Strauss partnered with a tailor named Jacob Davis to fulfill a need by miners to make a pair of pants that would last. Metal rivets were added to durable denim, and blue jeans were born. This was long before CRM technology existed; the *process* of Levi listening to and engaging with his customers is what created his success. He realized that his customers didn't need just tents and other dry goods, and what they desperately needed were long-lasting trousers. Through his relationships with his customers, he shifted his plans as market demands changed and as a result, became a great example of how CRM works. And he did it all without a laptop or a mobile device.

This is not to say that technology isn't a key component of the CRM process. No, a quick glance at the success of all of the CRM technology companies in the industry today will validate that. Whether it is tracking leads, upsell opportunities, or details of specific clients or prospective clients, technology aids greatly in *supporting* the CRM process and opens up new possibilities for building relationships with customers. But we should not lose sight of the fact that the technology only enhances the process by giving us the tools to deliver CRM – it does not replace the process. For my readers who are in an information security role, you understand that simply plugging in a technology to address, say, data loss prevention (DLP) doesn't equate to an effective DLP program. There are processes and data classification work that must be done prior to implementing the technology. Putting bad or incomplete data into a technology platform simply results in bad and incomplete data coming out of the system, albeit in an automated and efficient manner.

As a CISO, I have partnered with hundreds of sales executives in the information technology (IT) industry and have had a multitude of experiences in casually interacting with countless others. There are many who are very good at providing the services I need in a professional way, and I consider them to be an extension of my team. A few of them, perhaps three over my career, have teetered on the boundary of effective account representative and supportive friend. While these professionals dig deep into what I need to be successful, other representatives only contact me four times a year, always near the end of the fiscal quarter and definitely close to renewal time. This is not a relationship; it is a transaction. The difference between an effective rep and one that is only looking to close deals is the CRM process (not technology) they use. And if by chance a successful salesperson is indeed using technology to manage the relationship, it is behind the scenes and it is certainly not what makes them good at what they do. Instead, they are willing to take the time to get to know me, understand my needs, but most of all, build a relationship with me. This relies on a process for them in order to better gain my trust.

With all of the benefits and deep-seeded value of a proper CRM program, it is apparent why I chose this business domain as one of the benchmarks to assist me in delivering security services to an enterprise. I call it Security Relationship Management (SRM) and it incorporates the core CRM ideology, as well as many other aspects of marketing to deliver customized security to all corners of the business, as well as beyond the enterprise.

We should promote our security programs in such a way that makes it personal to whomever we are speaking with at a given time. After all, our audiences have different needs and different viewpoints on security, with varying frequencies of interactions that are needed. We cannot rely on a standard cadence of blanket security discussions across all the lines of business and functional areas and expect it to be maximally effective. The reality is that most of the message will likely be lost due to a possible perceived lack of direct relevancy.

Of course, we could assume for ourselves which parts of security would be relevant to a particular stakeholder, for example, our legal department. They will probably want to know more about how we can support privacy initiatives, record retention, or perhaps understand how our investigative processes could benefit their department. Wouldn't it be much more effective though to precisely cover how security relates to what they are thinking about, what is currently top of mind for them? In my career, I have found the only one way to accomplish this is to build a relationship with these entities and meet with them regularly to discuss – or better, listen to what is important to them. This interaction is not a "one and done" type of initiative. Relationships take constant nurturing, consistent communication, and a flexible, sometimes irregular, cadence to provide the support needed to strengthen them.

When I was in high school, I took Spanish language courses each of the four years I attended. After high school graduation, I took enough Spanish credits in college to be able to minor in Spanish. I even did a week-long trip to Mexico twenty-two years ago when my graduate school class studied at the Universidad Autónoma de Guadalajara. That was the last time I was immersed in the Spanish language. Today, I can barely order a meal in Spanish at a Mexican restaurant with any sort of assurance that I will get something reasonably close to what I ordered. I haven't used the language and as a result, it is lost somewhere in the deep recesses of my mind. The same holds true for SRM; it is a discipline that requires constant attention, or the message to the person will be lost over time, or at best, temporarily forgotten due to its irrelevancy to the current events for that stakeholder.

As a CISO, the breadth of our relationships can be rather extensive and quite daunting – future chapters of this book will help to validate this (as if you needed such validation; however, putting them on paper helps to appreciate the massive scope). We need to have solid interactions with all lines of business, every functional area, the executive team, our IT teams, our security team, the board of directors, as well as supplier partners (and some prospective supplier partners). Maintaining these relationships to promote security, customize it to the stakeholder, track interactions, and (importantly) receive feedback requires a significant investment of time. SRM is not a part-time job; it is one of the most important and time-consuming aspects of the CISO role. You are the voice of the cybersecurity program, and the person that stands between the company and its adversaries. Our corporate colleagues want to hear from you and learn how you can help them in their daily work. SRM aids in accomplishing this, but it requires meaningful organization if it is going to be successful.

Each of us maintains a variety of relationships in our lives. Family, spouse, significant other, our children, our pets, and a multitude of others. For the most part, we organize and maintain these connections without a considerable amount of formalized process work, aside from an occasional to-do list, to nurture them. But as we move outside our inner circle of family and friends, maintaining relationships across all of life's touchpoints can be daunting – particularly at work.

As CISOs, all too often we hear about the importance of aligning with the business, or perhaps frequently referred to as enabling the business. I completely agree with these mantras and believe it is the most effective way to deliver a cybersecurity program to a corporation. However, I've always longed for more discussion among our industry on precisely *how* to align with the business. It is my hope that this book on SRM offers some suggestions to do just that.

SRM is the formalized process for tracking all the relationships you have as a security leader. It includes the number and type of interactions, topics discussed, materials used, as well as key data points about each of your relationship connections. Later in this book, we'll dive into the mechanics of SRM; however, we should first explore some key concepts of marketing (slightly adjusted for security) to build a well-rounded background in order to apply them to your SRM process.

3

Marketing Concepts Re-Imagined

Let's get this out of the way early on: I am not a marketing professional. I know many people who have devoted their careers to the marketing field; their knowledge and contributions to the discipline far surpass anything I could put onto these pages. I studied marketing in business school with graduate courses that provided me with two things: 1) a comprehensive overview of marketing, enough to be professionally conversant in the field and 2) the knowledge that I don't know nearly as much on the topic as my marketing friends. But what I capture in this book is more than enough for our purposes here of applying general concepts to enhance our information security programs. Should you wish to learn more about marketing, look to those professionals who have gained mastery in the subject – I invite you to browse their books, presentations, and teachings. It really is a fascinating field, worthy of a more intense exploration.

Prior to diving into how security relationship management (SRM) can assist us in the delivery of our cybersecurity programs, an overview of some general marketing techniques will aid us in making our SRM program optimally effective. My goal is to slightly alter these marketing concepts to better fit an internal security program. CISOs require time to focus on the development and ongoing operations of their security programs; by leveraging already established processes (slightly altered for security) from our peer executives in marketing, we can more effectively, and more quickly, promote our security programs.

DOI: 10.1201/9781003597483-3

I know going into this idea (of applying marketing concepts to security) with a clear understanding that for many corporate cybersecurity teams, the notion of selling a security product or service to their company doesn't apply to them. For the majority of us, we serve in a functional capacity, supporting those areas of the business that do indeed sell various products and services to consumers. In many cases, the employees of our companies do not have a choice of products or suppliers when it comes to the cybersecurity services deployed to them because the CISO and their organization determines which ones to use and provides them. But the underpinnings of these marketing initiatives are so important, and with a little creative flexibility, the concepts themselves can serve as a foundation for what we as security professionals are charged with in providing to the company. It places us in a market-driven mindset, one that brings business topics to the forefront when we are creating our security programs – in an environment where markets relentlessly impact how companies respond to various market conditions. We should embrace this market mindset during our cybersecurity planning, without abandoning the foundational security concerns. Additionally, the concepts set up nicely the implementation of an SRM program (more on this later).

As I dove into the individual marketing disciplines over the years, I realized that many of the components of a particular marketing process could help me with delivering security services, but other components within the same process just couldn't be applied – the creative reach in making them work left me feeling as if I was attempting to put a square peg into a round hole. It just didn't work, so I abandoned that specific step. But in no way does my excluding a step mean it isn't important for other marketing endeavors.

Lastly, I choose the words "creative flexibility" very carefully. The original intent of a specific marketing concept could be vastly different from how I apply it in this book. My readers who are marketing professionals may no doubt roll their eyes in disbelief at my translation of their concept to help fit it into delivering security services. However, that is what I did; I started with the general theme of the process and using some creative flexibility as a result of two decades as a CISO, applied it to cybersecurity. Please consider my translation of your concepts as flattery and a respectful nod to the important work you do every day.

Getting Started

At its most basic level, marketing is about influencing and creating demand. For cybersecurity professionals, influencing is already a large part of what we do every day. We attempt to influence appropriate cyber behaviors that contribute to protecting the company. And even though I believe that more people are curious about information security today than ever before, and how they can protect their company (and themselves) online, the challenging part of security still remains in creating a demand for it. How do we get demand from our employees for enhanced security, rather than it just being looked upon as a mere necessity of doing business in today's connected world? Without a plan to tailor our cybersecurity initiatives so that the consumers of our services feel their needs are being met, or at the very least, realize those services help to serve a greater good, creating demand will continue to be a struggle.

I start with some underlying marketing concepts and draw out those components that will enhance my relationship with the person, understanding what is top of mind for them, and then tailoring my services to meet their needs. The first step is a complete evaluation of the services I provide through the use of an industry concept called the Four P's. Simply stated, they are product, price, place, and promotion, and using this process helps me to understand how I will build and deliver my services.

Product

Determining which services I provide is based on the risks to the enterprise. Because each corporation is different, with varying levels of risks, affordability, risk appetite, and threat posture, our security services will be vastly different from company to company – as they should be. These attributes are typically drawn out using an ongoing cybersecurity program evaluation, risk assessments, outputs from our security technologies, and through close interactions from those who consume our services – particularly corporate leadership who fund our efforts. Through these mechanisms, we ensure that our services actually fulfill a demand and can highlight a balance between the goals the business is trying to achieve, with directed risk mitigating strategies that seek to enable business goals.

Our security products may take the form of an actual digital product on an endpoint, or perhaps a service, that seeks to address a specific cybersecurity concern. In building out our product (or service) strategy, providing insights into their features and capabilities helps to validate that they fulfill a need, namely, mitigating a risk. Additionally, it highlights the problem (or the challenge) that the product/service aims to address, or potentially a regulatory requirement that is met in providing the solution.

Quality is another important aspect of your product portfolio choices. Is there a difference in quality between what your internal team can provide and an external product/service provider's capability? Does the cost associated to insource a capability justify the quality of the service? If we choose to go with a third-party supplier, several other evaluations must be performed to ensure that the CISO is purchasing the highest quality offering, at a reasonable cost (this doesn't always mean the most inexpensive price). There are more cybersecurity vendors today than ever before and to determine the best fit for your organization requires deep analysis of the vendor landscape.

Lastly, cybersecurity doesn't exist in a vacuum; therefore, our determination of products and services must include an investigation into how they could (potentially) negatively impact other portions of the business. One example that immediately comes to mind is any latency or performance impacts resulting from the security product or service (whether it is internally or externally sourced). Through strong relationships across the business, you'll be in a better position to highlight those impacts.

As previously mentioned, it is understood that many times the end user or customer of our services do not have a choice of product selection when it comes to an internal cybersecurity program. But leveraging this market-driven type of product focus will optimize your product/service portfolio. With some attention to the "product" aspect of our services, we ensure they are fulfilling a demand, which, in turn, differentiates our cybersecurity program from other programs that may be vastly different from company to company.

Price

At this point in the book, you will start to notice the creative flexibility I referred to earlier. Of course, the price component in the original Four P's concept is directed to the competitive pricing of a product or service in the market. As a CISO delivering an internal program, you are not seeking to create a profit by selling your cybersecurity product or service within your own company. Instead, I tweaked the concept a bit to apply it to cybersecurity services for an internal corporate function. Important to note though is that if your company does indeed provide cybersecurity products and services to external customers at a competitive price, one may want to leverage the Four P's model as it was originally intended. My application of the price concept focuses on two components: allocation and budgetary funding.

Depending on how your company allocates security costs internally, if it allocates them at all, the "price" of your service could be impacted. The actual service you deliver may often be the same, but the timing for the delivery of the service may have variability due to tight schedules by the business. For example, let's imagine that a functional area requests a forensic capture of a computer, but they need the service faster than the standard service level objective. The service remains consistent, but the accelerated timing introduces differences in the costs associated with providing the service. These costs can be in the form of more internal labor costs dedicated to the service request, or possibly outsourcing the request to meet deadlines. If your company allocates security costs based on the discrete services provided, it is reasonable to identify this price variability in the form of different service levels, perhaps standard, gold, and platinum. The faster the delivery, the higher the cost. Even for companies that choose not to allocate security costs based on discrete services, the activity of outlining the cost differences in service levels can prove valuable to the CISO for things like capacity management and resource planning. Simply put, measuring how often your lines of business accelerate your services beyond the standard service levels (for which you are not recouping costs) provides good insights into future planning. If trending data shows the majority of business units require faster delivery of security services, this could be a good indicator that the service levels need to be adjusted – or that additional resources are required to accelerate the service delivery.

Perhaps the largest driver for the implementation and ongoing maintenance of a security program revolves around funding the cybersecurity budget. Without the appropriate funding for evolving security efforts that address likely risks, the program will not be successful. Many may consider a material lack of security funding to be detrimental (and I would agree), but the inverse is true as well: too much security funding in the wrong security domains can be just as unfavorable. So how does a CISO deliver the appropriate request for security funding to executives? To outline all of the incubators that lead to an effective security budget proposal is another book in and of itself. For the purposes of our time together now, it is only to say that ensuring a comprehensive study and resulting financial plan for the security strategy is vital;

leveraging the price attribute of the Four P's has helped me in my work to accomplish this task. It may not be the intent of the original model, but with some creative flexibility, the Four P's process can help the CISO to determine some of the cost-related aspects of their security program.

Whether your security costs are allocated or not, pricing work is valuable insofar as it provides deep analysis for how prices (or costs in a non-allocated model) are derived. One aspect of determining price is looking at competitor pricing. While not a true competitor to your internal information security organization, a third-party vendor's cybersecurity offerings may often initiate questions from leadership on why we would not consider them for our program over an internally provided service – particularly if that service is cheaper with the external entity. For example, if a CISO is looking to build an internal capability for monitoring and detecting threats, they should consider other participants in the cybersecurity market. Building an internal capability for this service can be incredibly expensive. At the same time, there are established and emerging entrants (external considerations) who provide this service leveraging massive economies of scale with strong geographic placements that make their offerings worth a strong evaluation. Additionally, these external entrants may have established cybersecurity capabilities that could surpass what an internal security organization could build in the short term.

Through a deep analysis of the price/cost between external and internal capabilities, the financial projections are now steeped in a more complete view. There are more than just labor costs associated with building a 24×7×365 monitoring function, costs which may vary dependent on many factors. Several other fixed costs such as facility (heating/cooling, security) and employee amenity costs (break areas, food, beverages) may need to be considered for those companies that reduce these services outside of normal office hours. The decision to outsource or build the capability in-house is up to each individual corporation; however, the deep cost analysis performed can (and should) be used by every

CISO. Let's look at a few more financial areas where the Four P's model can assist us.

If one of your specific cybersecurity services has extremely high demand across your enterprise, funding the service appropriately to meet that demand is the next logical step. But what if there is a service that you want to include in your service catalog that isn't provided today? Because funding the service isn't influenced by current high demand, along with the uncertainty of the future demand for the new service, an effective strategy may be to fund it incrementally. The CISO may identify a demand trend in one of the lines of business for a specific service through their SRM work. Yet, it still may be unknown if the service will be widely adopted across the entire enterprise. By starting small, a service could be added to the service catalog and monitored to see how enterprise demand for it follows.

At one of the companies I worked for early in my career, we thought it would be beneficial to include a digital forensic service in our capabilities. We knew that one of the corporate divisions would heavily consume the service because of the unique work they performed, but we were unclear if the rest of the divisions would find value in such a service. We started small by purchasing a bit of hardware and a few software licenses. We sent three members of our security team to forensic training, specific to the software we chose, to learn how to capture and analyze forensic images. We found in the following year that more divisions and functional areas were interested in the service, so we expanded our capability.

It isn't unusual for a cybersecurity organization to deploy "security bundles" to project work, particularly those projects that cross the corporate network boundary. Requirements such as intrusion prevention, next-generation firewalls, endpoint protection, DLP, and many more are collected into a security bundle. For the project to commence, requirements must be met by deploying the security bundle at the boundary. And because projects vary in size and scope, it is a good idea to identify the costs associated with security bundles for all sizes. For example, a project of a small size can expect to pay (or absorb) an estimated amount. Larger projects would naturally generate greater security costs for the bundle. Such cost transparency helps project teams to anticipate costs, while assisting the CISO with demand management and resource allocation.

As an internal function, our cybersecurity teams may not naturally focus on pricing since the costs are often either allocated across the enterprise or absorbed in a centralized overhead budget for security. Depending on the industry you are in, rarely does an internal security team provide direct or immediate financial benefits to top-line revenue growth. However, leveraging the price portion of the Four P's model details the price/cost underpinnings that are important for helping to increase bottom-line efficiencies.

Place

Deep planning to determine how we will deliver our security services ensures we place the needed services appropriately. As a corporate function, especially if the security organization operates in a centralized shared service model (one organization providing services to the entire enterprise rather than several disparate teams addressing security within the same company), the services will most likely be delivered to the whole of the enterprise. Yet, some variability may occur in delivery due to unique needs of a specific team or geographic region.

To obtain maximum effectiveness of your services, some work needs to be done in determining how you and your team will deliver the services, and through which channels. Borrowing heavily from marketing concepts, the CISO ensures services are placed so that the applicable corporate stakeholders have access to them, when they need them. Distribution channels make up the process of how we get these services to our customers (in the case of cybersecurity, to our employees and leaders).

Even though your security organization may be operating in a functional model that often occurs behind the scenes, detailing cybersecurity distribution processes are incredibly important to ensure consistency.

Generally, a primary channel for obtaining service requests and the subsequent delivery of those cybersecurity services is through a Help Desk or Service Desk ticket. The employee submits a service request ticket that is tracked until completion. Details on what service is requested, the history of any problems the user may be experiencing, and which member of the cyber team is addressing the request may be typical elements of a service request. This represents a value-added mechanism for ensuring that each request is routed to the correct service delivery team, that it is tracked, and that it is auditable. But often there may be cases in which this service distribution process is bypassed; therefore, other alternatives should be recognized.

Consider this analogy to a possible alternate distribution channel: the Executive IT Support or White Glove Support function for the highest levels of leadership within a company. Rather than a corporate executive putting in the standard ticket for IT services (as is the normal process), a small group of IT professionals may be assigned as Executive Support. With just a phone call, text, or email, the executive can have immediate, direct, and in-person support for their technology needs.

There may be cases when an employee has a question about cybersecurity that may ultimately result in a service being required by the security team. The employee may reach out to an acquaintance in the security organization asking for their help, again bypassing the standard ticketing distribution process.

It is not uncommon in very large enterprises to have a role serving as an intermediary between the lines of business/functional areas and the core cybersecurity team. Perhaps called Business Information Security Officer (BISO) or Cyber Risk Partner, this role sits with the assigned line of business, attends their meetings, and understands their roadmaps for upcoming initiatives and strategies. The role then brings this information back to the information security team to better meet and anticipate future service needs. It is not unusual then that the employees from that line of business will feel more comfortable reaching out to their BISO for facilitating their cybersecurity service needs, rather than initiating a service desk ticket.

Perhaps there are geographic considerations that need to be examined to ensure effective distribution of your cybersecurity services? Different time zones, availability of resources, or unique regional needs, all could lead to distribution implications. A country manager may feel more comfortable with requesting services from their onsite security representative. These considerations should be understood when building out your distribution channels.

We all need assistance from time to time, and marketing your security program is no different. Very early in my professional career, I spent two years in the strategic alliances division of a large company. Our goal was to develop partnerships with other companies that could add value to both companies by bringing awareness to one another's capabilities to our collective customers. These alliances are a great way to expand the company's message to more audiences, and it fills in the gaps that may exist in your standard distribution processes. Several years after my time in the strategic alliances organization, when my transition to cybersecurity had occurred, I reached back to those times to see how the concept of strategic alliances could be adapted (in a formalized way) for distributing my cybersecurity services. This is what I found.

Aligning my program with other organizations in a shared effort created a much richer benefit than if I were to go at it alone. For example, several years ago when seeking budget from the CFO for a technology to effectively remove administrative rights on endpoints, I didn't lead with the security benefits. Instead, I partnered with the vice president of infrastructure and the director of our desktop support team. We communicated the benefits of reduced labor costs for IT personnel having to re-image a computer due to an end user's misconfiguration of the asset. The infrastructure executive spoke about the impacts to the computing infrastructure due to employees having administrative rights. Only then did I speak to the security benefits of removing administrative rights and the reduction in risk of malware being automatically executed on an employee's machine. This strategic alliance between the three teams produced a common front, each with different goals, but all enjoying strong benefits from purchasing the technology solution.

Similarly, at a different company many years ago, I presented my initial cybersecurity strategy and associated budget. I knew at the time that many of the security initiatives I was promoting would cause a strain to the company's IT organization, particularly within the infrastructure team. They would have to build out the environment, test it, deploy it, and provide ongoing support for many of the initiatives I was requesting. As a result, in my labor request there was a line item for two full-time infrastructure employees. The CFO asked if I was building out a small infrastructure team within my security organization, and when I told him that I was not, he was confused. I explained that much of the work in my strategy fell on the backs of the infrastructure team and my labor request was to add resources to *their* team. This was unprecedented at the time for this company; one organization requesting labor budget for another organization just wasn't done. But the strategic alliance I created with the leader of the infrastructure team allowed for the promotion of my strategy, while meeting the needs of their strategy as well.

Ever since I experienced these examples from early in my career, I have continuously sought out strategic partnerships with all areas of the business to promote and distribute my cybersecurity services. Most recently, the data team within my company was conducting a data analytics week with representatives from all over the company. Employees flew to the corporate headquarters from all corners of the globe to deliver informational sessions to the employees. Through a strong alliance between our two groups, the leader of that team and I partnered on the event and my security team presented two keynotes on our cybersecurity services (specifically around data protection) at the event. Strategic alliances have enhanced my information security distribution channels and remain an effective part of my placement strategy.

The point of leveraging marketing processes is to ensure that as we build out our cyber strategies, we place our services so that the consumers of them have access to them, when they need them, and in a place where they will be likely to search for them.

Another positive outcome of understanding our channels is to more efficiently manage our "inventory" of cybersecurity products and services. Formalized channels lead to better demand and usage data, data that can be incredibly valuable when determining our needed inventory of licenses for our security solutions, or perhaps understanding our cybersecurity staffing needs to ensure we are meeting the demand for our services.

To maximally deliver the services in such a way, a clear understanding of what services your security team provides is needed. I found that developing a cybersecurity service catalog is a wonderful way to document our security services, what employees can expect from our services, and important to the "place" aspect of the Four P's, how to order the service. The catalog is written for the potential "customers" of your services using easy-to-understand, non-technical language, and is accessible in a conspicuous area of your company's internal website.

The components of your cybersecurity service catalog may vary, but could include:

- **Short Description**: a quick explanation of the service;
- **Long Description**: a detailed description of the service, what a consumer can expect, and the outputs/results of the service when completed;
- **Costs**: detailed cost breakdown for the service (even if costs may be allocated rather than charged by service);

- **Reliant Services**: those services that cybersecurity does not provide, but are essential in delivering their security services, i.e., network connectivity and cloud infrastructures;
- **How to Order**: a description of exactly how to order the service;
- **Service Level Objectives**: how long it takes to perform the service, attributes of the completed service, and related objectives.

For CISO readers, you may correctly realize that many of the services your organization delivers do not require a request; they are simply delivered to the corporation behind the scenes. Regardless, the service catalog highlights your services and provides employees with information on the services. Even for those capabilities that do not require a service request from a user, the value of creating a security catalog helps with branding your cybersecurity organization, a concept that we'll explore together in the next section.

Promotion

If security is primarily thought of as a back-office initiative, is there truly a need to advertise and promote the security program across the enterprise? Many agree that promoting security is important because we want our employees to understand the importance of cybersecurity and the risks that are ever present. But the promotion aspect of this model is not directed to promoting the importance of security; instead, it is the promotion of the cybersecurity program and its services themselves. Let's explore some things we can do to promote our security program, as well as some ideas to advance awareness of what we deliver – not solely the importance of the security message itself.

With just a quick glance, a good brand will elicit thoughts about the products and services associated with it. Merely hearing the name of the brand can produce thoughts as well. The goal of any good brand management program is to connect with customers (as well as potential customers).

I believe that it is important for a cybersecurity organization to put in the effort to develop a strong brand to connect with the employees of their company, which, hopefully, will keep security efforts top of mind for them. An effective brand management program for cybersecurity has two benefits: 1) builds awareness around the security topics themselves: i.e., phishing and weak passwords and 2) promotes the capabilities of the cybersecurity organization for all corners of the business. As I outline a few tips that I've found successful, you will find that the two benefits are tightly intertwined, but as previously mentioned, the primary focus for cybersecurity branding is to bring awareness to your organization's security capabilities. In turn, this can lead to more awareness of the security topics.

Historically, information security teams operated as a back-office function with a healthy dose of ambiguity as to what the organization did and how they performed their services. But as cybersecurity emerged more and more as a frequent, critical business imperative, increased visibility and curiosity into the security function followed. Your employees want to know more about how your information security organization supports them, as well as how they can protect themselves from the consistent onslaught of digital threats. A cybersecurity program is part of the business and should promote itself effectively so that it is viewed as such.

I start with a creating a simple visual logo to represent the information security program. The logos I have developed were not a wide departure from the corporation's logo; usually it is just the company logo with the name of the cybersecurity organization underneath it, perhaps a small icon to accompany the cyber team's name. Using your corporate color schemes is another important aspect of building a logo. After all, your corporate marketing and communication teams aim to accomplish the same thing as your security branding strategy: a common theme and brand to make a connection with the company and its consumers. Leveraging vastly different logos and color schemes from your corporate brand only produces confusion among those with whom we are all trying to connect.

Armed with a logo and consistent color theme, every message, awareness article, security poster, and presentation that is delivered to the enterprise from the cybersecurity organization contains the new branding design. The goal is to provide our employees with immediate recognition that the message is coming from their information security organization and produces instant identification of the brand.

For whatever the reason (although a quick Internet search of *mimetic desire* is well worth the time), most everyone enjoys receiving free stuff. I was once at a security conference and was amazed to see a line of at least seventy-five people waiting patiently to pick up a pair of cheap, neon green, plastic sunglasses etched with a cybersecurity company's logo. We love free things, and I capitalize on this by purchasing a variety of items adorned with the newly created information security logo mentioned earlier. Shirts, tumblers, mobile device chargers, web cam covers, and many more items are used to promote the cybersecurity organization. They can be passed out at security training events, as raffle items, or perhaps to those who properly report a potential security risk. When the employee proudly wears the shirt or jacket, their peers frequently asked how they might be able to obtain one too. I have found across multiple corporations that I have been part of that these items are highly coveted by employees, and to this day, remain an important part of branding my cybersecurity organization.

But logos and free gear are just a foundation on which to build. Security organizations must effectively brand their program capabilities leveraging other mechanisms as well. At least once a year, it is a great idea for CISOs to plan and conduct a cybersecurity open house. Here's how it works:

Prior to the open house event, the information security organization works with their security vendors to obtain small giveaway items, such as pens, cups, and water bottles, to provide to employees at the event. A large conference room, common area, or perhaps a section in the company's lunchroom is reserved. Tables are set up with large monitors into several stations, with each station representing a capability the security team provides. Example stations could include endpoint detection, vulnerability management, security awareness, and so on, each one staffed by a member of the cybersecurity team. Employees browse through the stations at their own pace, learning about the different capabilities from the experts. An information security team member shows the technology that is used and answers any questions the employees may have. A table is set up with all of the free gear that the vendors provided for employees to grab. To enhance the open house even more, food can be brought in, such as an ice cream sundae station or slices of pizza. Open house events are a wonderful way to promote your security program, highlight cybersecurity as part of the business, and effectively brand your organization.

Promoting the information security team doesn't need to be limited to large events like an open house; I believe cybersecurity occurs in conversations. One method I found impactful is the 5/55 technique for promoting the security program. Working with my peer leaders from other parts of the business, I'll offer to speak on cybersecurity at one of their team leadership meetings for five minutes. I'll do a quick overview of a cybersecurity topic like phishing and data loss, I'll talk about how the cybersecurity organization works to protect them, and then I'll explain how the team can connect with us. It is a quick hit that keeps security top of mind for the audience and gives them a chance to ask questions. But the real value in this method is not the five minutes when I am speaking; it is the fifty-five minutes that I spend listening to the other discussion points from the team during their meeting. Hearing their strategies, pain points hindering execution, and other updates really helps me in determining how best I can support them. This is an example of getting as close to the customer as possible and it works great.

Branding the cybersecurity program helps to build connections with the consumers of your services. With a bit of planning and a little effort, your cybersecurity branding strategy will provide you with actionable information to showcase, as well as enhance, your security program.

At this stage of the book, it is an appropriate time to transition from broad marketing concepts to specific processes that transition us closer to executing SRM. I believe the most important of these processes is consumer segmentation. This is where we begin to differentiate between the consumers of our services, and it's the topic we will cover in the next chapter.

4

Segmentation (Not the Network Kind)

Recently I received marketing material in my mailbox from a company that specializes in swimming pool maintenance. The service is very complete; the company monitors and adjusts the chemicals in the pool, cleans the filtration system, skims the debris, and routinely checks the water level. The company has great reviews from its customers, all who consistently sing their praises on the excellent quality of service and friendly staff. The price for their service is extremely competitive with a wonderful value for the service provided. The marketing material was visually appealing and very engaging. I thought it was a fantastic marketing campaign for what appears to be a great company.

I don't own a swimming pool.

To effectively target and engage your customers, both current and prospective, you need to know your customer, and what is important to them. Generic marketing messages to your employees on the capabilities of your security program have some benefit, but they will never be as effective as targeted communications designed specifically for what is top of mind for each employee. Obviously, CISOs cannot create individual messages for every user, but there is a way to obtain more precision in targeting their needs through a marketing concept called segmentation.

Wouldn't it be much easier, and more effective, to divide all of the employees of your company (as well as external stakeholders) into manageable sub-groups based on similar attributes? This is the core principle behind segmentation that allows CISOs to target

DOI: 10.1201/9781003597483-4

their message for maximum applicability. Segmenting your core audience provides you with opportunities to develop unique cybersecurity services and messages that directly meet the demands of a particular consumer segment within your enterprise. Your message will be more personal to what is top of mind for that specific segment, and it will be delivered through their desired communication channel. It removes vague messaging and opens up communication channels in which the segment may be more responsive. The principle of segmentation also allows you to test new cybersecurity services for those segments that may be more security-forward than others and introduces a new feedback loop into your cybersecurity organization. The work in segmenting is intensive, but the benefits are extremely valuable.

When I began my work on segmenting my cybersecurity audience, I used core marketing segmentation principles to get me started. While they have evolved to (arguably) include more, the basic segments are geographic, demographic, psychographic, and behavioral. Just thinking about these general attributes was enough for me to brainstorm on other unique characteristics that could lead to an effective segmentation strategy for my security program.

I wanted to achieve segments that allowed for targeted messaging, but not segments that were overly small or particular. For example, one popular segment I use for my cybersecurity program is the legal department within my company. I stop there because dividing the segments into smaller groups within legal such as intellectual property, litigation, ethics, and compliance, as well as other various types of legal teams, does not produce more value for me, and often introduces increased administrative work in maintaining them – so I stick with just a general segment of legal to cover all of their teams. This will become clearer as we discuss the administrative tasks associated with security relationship management later in this book.

In addition, I am flexible in developing and maintaining my segments. Business evolves, organizations shift, and strategies change, which requires that my segmentation strategy be fluid and can easily adapt to these changes. On more than one occasion I had to tweak my segmentation program due to a reorganization of the company – in one example, transitioning from nine business units to just three operating segments. The combined business units into operating segments created new strategies that served the segment over the business unit, as well as altered the leadership makeup of the groups. Like cybersecurity, segmentation is not a one and done, project-type of initiative – it needs frequent nurturing.

The way that I have thought about segmentation over my career is that the top leader for each business unit, sector, division, functional area (or insert any other applicable name here) thinks of their organization as a "mini-company" – one in which they are a CEO. For large corporations, those units may even have a dedicated CFO, CIO, or other senior executive type of role reporting to the unit's top leader. These units contribute to the overall success of the company and to do so, they run their area like a business in and of itself. Effective segmentation has allowed me to deliver specialized content and consultation to each of those leaders with this approach in mind. Even more valuable is when I can compare data from that unique unit to the data from other units within the corporation.

Lastly, there may be cases in which two or more segments overlap. For example, if one of your segments is manufacturing, and one of the company's manufacturing plants is located in Ireland, that plant could be part of both the manufacturing and international segments. This is natural, and the variable information between the segments could add insights potentially missed if only identified as a standalone segment (e.g., how the manufacturing facility compares to other functions in the international segment regarding the output of product units). I find it helpful to identify these areas of overlap within my segmentation work and depending on how you organize your segment administration work, potentially hyperlinking between the two.

The marketing concept of consumer segmentation has really helped me in my journey as a CISO, probably more than any other of the concepts that I borrowed from the marketing industry. It personalized my cybersecurity program to fit the needs of the person or team I was talking with and, subsequently, helped to increase their frequency of involving my team in their initiatives. In order to get you started on your segmentation journey, what follows in the next chapter are some suggested segments that I found useful. Of course, you know your business better than anyone, so customizing what I offer in the following pages is not only recommended, but also needed in order for your program to be successful.

5

Segmentation Suggestions

Let's get started with building your segmentation strategy by suggesting a few segments that have performed well for me over my career. Perhaps they could provide value for you as well, or at the very least, serve as a primer to begin thinking about the potential ways you can segment your target audience.

Suggested Segments

Corporation

There are times when a CISO brings an enterprise view to the work they do; the target "market" is the entire company, a segment which isn't a large departure from what CISOs do today. The work that we do and the metrics we collect are all part of our daily service delivery to the company as a whole. You may often send awareness articles to the entire corporation, or maybe deliver phishing campaigns to all employees, as well as provide annual security compliance training. Information security policies, organizational budgeting, and other endeavors often see the company through an enterprise lens as well. Whatever the initiative, the group you are targeting is the entire company within the specific industry in which you operate. Because of this, the corporate segment is very familiar to CISOs and should remain as part of the segmentation strategy – not as a segment of a larger group, but the overall group itself and one that requires a degree of specialization that differentiates itself from other companies.

DOI: 10.1201/9781003597483-5

Board of Directors

The relationship between the board of directors (as well as board committees) and the CISO is one of the most important ones a security executive can nurture. There are books and volumes of online publications that discuss the relationship between the CISO and board members, so I won't repeat what can be found there. Instead, I only suggest that the board is one of the most significant segments a CISO can target, and as a result, should be an important segment in your strategy. Each director brings experiences from previous companies that they have worked for – perhaps across multiple industries. Many times, your board members serve on other boards that have their own unique cybersecurity programs. Understanding each of the director's background helps to frame their experiences with your own discussions with them. Additionally, I often connect with CISOs of companies where one of my board members serves on their board as well to exchange notes and insights about the director – the questions they ask, the topics of most interest for them, and many more.

If you are lucky enough to have a board that leverages your expertise using a variety of interactions outside the typical quarterly board update, tracking these interactions becomes even more critical. Or if minimal interactions occur between you and the board, segmentation unlocks more information that could potentially persuade the board or committee to hear about cybersecurity more frequently from the CISO. No matter what level of relationship, the board is a very important segment to outline in your strategy.

Executive Team

The Chief Executive Officer (and his or her direct reports) is a segment that requires special attention, and in a manner vastly different from the other segments you identify. In many ways, they direct the cybersecurity program and its strategic direction (and security culture). Most often, these are the executives who seed your program with funding. Targeting this segment allows the CISO to engage more closely with them. In doing so, you could evaluate and document the behavioral aspects of the leadership team, i.e., which member(s) of the team are more supportive of the security program, which executive seems to hold more power in the decision-making process, and other qualitative information. Similar to the corporation segment, working with the executive team should not be unfamiliar to you, but our goal in segmentation is to formally build a process in which you can identify the unique characteristics with each of them.

Cybersecurity Governance Committee

Many companies leverage a committee to provide governance and oversight of the information security program. Made up of executives from varying parts of the company, the committee meets on a regular basis (perhaps quarterly) to help set the direction, allocate resources, and arbitrate conflicting business priorities that require support from the cybersecurity team. This is a crucial segment that requires cultivating, particularly outside of the regular meetings. Interacting with the individual members of this group will help highlight important topics that can seed agenda topics for upcoming committee meetings.

Top Cybersecurity Proponents

I always include a target "market" in my segmenting strategy for the group of people across the company that support cybersecurity initiatives at elevated levels of involvement. Typically, this segment is made up of members of the Information Security Champions program, a formalized, voluntary group of individuals who help spread the word about proper cybersecurity practices within the company. In return, they have access to specialized security training, recognition for their efforts, or perhaps small tokens of gratitude such as corporate branded information security gear. This segment deserves special consideration when addressing cybersecurity, and providing them a segment in the strategy helps to target your message, which, in turn, is spread throughout the enterprise by these dedicated employees. And while they don't deliver cybersecurity services, they expand the demand of your services and provide vital feedback into the security needs of the enterprise, as well as the satisfaction levels of your cybersecurity program.

Other Committees

Apart from my duties as a CISO, I have often served on various, non-security committees. At two of the companies I've worked, I served on the company's disclosure committee. We would meet quarterly to discuss the upcoming reporting of financial (and other) disclosures, and it is a wonderful way to keep informed of company performance and key strategic initiatives. And with the 2023 U.S. Securities and Exchange Commission rules on cybersecurity disclosure, the security discussion within this committee has increased significantly. Committees such as these are important enough for the cybersecurity program that I always include them in my segmentation work.

Business Unit

Call it a business unit, an operating segment, sector, division, or a variety of other names, each one of these entities deserves a place in your segment strategy. They drive the business and are the backbone for much of what the company achieves. You will want to ensure that each one of these units has a dedicated segment assigned to it so that you can understand as much as possible about their strategies and operations. Above all else, you'll get a better understanding of how cybersecurity can be woven into what they seek to achieve each day. For many larger corporations, the CISO may receive support in maintaining these segments through the BISO, Cyber Risk Partner, or similar role. Even with this assistance, segmenting these groups in the CISO segmentation strategy provides the security executive with detailed information to maintain strong relationships.

Manufacturing

This has always been one of my favorite segments to work with because I obtain first-hand knowledge of how raw materials are transformed into the products the company delivers to its customers. For me, the type of product doesn't matter; the manufacturing processes are fascinating to learn about and explore. For this segment, you'll want to understand all aspects of production including any robotics that are used, machinery vendors, downtime between shifts, maintenance windows, and various metrics associated with manufacturing. For example, it is wise to understand inventory details (both raw materials and finished product levels), as well as daily product unit output, to name just two, both of which can contribute to understanding impact to a potential cybersecurity breach.

Through a deep relationship with the manufacturing team, you will also get a better understanding of how some of your security services, i.e., vulnerability scanning or penetration testing, could potentially negatively impact operations.

Distribution

Any function that moves a finished product from your manufacturing plant to the customer (and all points in between) is an important one to include in your segmentation work. Transportation, logistics, shipping, and whatever organizations are aligned to this work can be included in your distribution segment. Through your work with this contributor to the business, you will achieve a better understanding of which shipping vendors your company uses, its wholesalers, as well as key retailers.

All of these elements can be directly related to cybersecurity with regard to safeguarding the supply chain.

International

If your company has operations across multiple countries, no doubt your CISO duties have included many international interactions – why not formalize them in your segmentation strategy? For some companies, there may be an international division, specific country divisions, or perhaps country leaders responsible for a specific region. These are important stakeholders to have an ongoing relationship with due to a few reasons. Varying compliance laws between countries is one that quickly comes to mind. There are cultural variables that also come into play, each with different risk tolerances.

Another critical reason for closely aligning with international segments is for those CISOs who have a portion of their cybersecurity operations overseas. A balance needs to be achieved between those resources being dedicated to your centralized cybersecurity organization (if applicable) and those providing local regional support at the same time. Because the cybersecurity resources are local, they may be increasingly leveraged by the country manager to support their own initiatives. This is not necessarily a bad thing, but there needs to be a balance in local hands-on support and the federation of the corporate information security shared service. And this is better achieved through close and frequent interactions between the CISO and the country manager.

Human Resources/Talent Management

The relationship with your Human Resources (HR) department is a very important and a very reciprocal one. Supporting one another through recruiting and hiring talent for the cybersecurity team, standardizing security job roles, and conducting investigations are just a few ways these two departments assist one another. Because the people aspect of cybersecurity is so critical to an effective program, the HR segment should be near the top of your segmentation planning.

Finance

This segment could include more than just the finance team; I have usually lumped in areas such as accounting and procurement into it as well. The relationship with the teams that assist with budgeting and purchasing of security products and services will be more effective if it is maintained throughout the year, and not solely during budgeting season. For example, one of my relationships with finance many years ago resulted in proactively creating time codes to track labor hours toward incident response work, so that if a material incident took place, we could calculate the cost of responding to such an incident much more effectively. Additionally, when I needed to do some financial forensics to track purchases and owners of non-IT supported initiatives, the finance team provided wonderful support. But this relationship should not be limited to cybersecurity finance, the CISO can also learn a great deal about the entire financial posture of the company as well. Finance is always on my segmentation list.

Sales

In my experience, the individuals in this segment are always great to work with; their excitement about the company's products and services is palpable. They exude personality and create the linkages between all the corporate functions and the customer. Your sales representatives are the face of your company and communicate the company's values to the consumer – I am sure there is a cybersecurity opportunity in there somewhere. By including them in your segmentation strategy, you'll be in a better position to identify those opportunities and capitalize on them.

Information Technology (IT)

This segment is a no brainer for CISOs; for many in the role, it is the home organization in which they reside. For those who report outside of IT, the importance of the relationship between cybersecurity and IT isn't diminished. It makes sense then that the relationships between the CISO and across all of IT are significant. This is where the execution of technology solutions happens, with many cybersecurity initiatives falling on the backs of the larger IT organization. While I have never tried it in my segmentation strategies, it may make sense for your strategy to slice up the IT organization into multiple segments. Perhaps things like enterprise architecture, infrastructure, network, and cloud, depending on the size of the IT team. If your company's IT department is small to medium size, the broad IT segmentation may be enough for your plan. Whichever strategy you choose, the IT segmentation is a critical one for the success of your program.

Digital & eCommerce

It isn't unusual for some digital and eCommerce teams to be organizationally separate from the IT department. If these functions are indeed part of the IT organization, putting them in your IT segment may be adequate. However, if they are organizationally separate, or the work they do is such a large part of the strategic direction for the corporation, it is wise to create a segment dedicated to these teams. Besides compliance interests, CISOs can leverage this segment to understand other security domains such as application security, security awareness, and many others, related to the digital initiatives for these teams who strive to produce omnichannel mechanisms for revenue.

Communications/Investor Relations

A great relationship with the corporate communications department can greatly increase the effectiveness of your security program. Their role is to manage the message both inside and outside the corporation; having cybersecurity top of mind for them could help with the timing and frequency of the CISOs' security messaging.

Legal (Including Outside Counsel)

I cannot think of one CISO role that I held in which the legal team was absent from my segmentation strategy – the relationship is too critical not to include it. The two functions blend well together because of similar concerns they share across multiple disciplines: data protection, privacy, and a host of others. Just as important is the relationship with outside counsel, particularly when it comes to incident response, or perhaps better stated, responses conducted under privilege. Segmenting this function and the subsequent relationships within it will clarify areas in which the CISO and the legal team can support one another, ideally before they become major issues.

Internal Audit

Leveraging internal audit can be quite useful for CISOs to help identify risks across the company, as well as helping to promote your recommendations for addressing those risks. Most internal audit functions have the ear of senior leadership and the board; their reports are critically reviewed and discussed by the highest levels of corporate governance. Through a strong relationship with audit, your cybersecurity concerns can be examined by a separate organization within the company. Just as important is your relationship with the company's external auditor, one that should be cultivated regularly and not just before board or audit committee meetings. Segment this group and you will be in a better position to support one another in many of your shared initiatives.

Top Cybersecurity Vendors

Segmentation, and the resulting security relationship management program, should never be limited to inside your company. We've explored a few examples already of relationships not within your corporate walls, i.e., outside counsel and external auditor. Building off those examples, I always create a segment for my program's top cybersecurity vendors. These are the partners that I consider to be extensions of my team. A very clear example would be a managed security service provider (MSSP) if the company uses one. There are customer reviews of MSSP offerings (even from the same service provider) that are exemplary, while other reviews are sometimes negative. The platform and service may be consistent, so why the variability in customer satisfaction? In my experience, customer satisfaction depends largely on the relationship between the MSSP and the security team. And there is no better way to enhance that relationship and resulting performance than by segmenting these vendors to understand the needs of both parties, in order to build a solid, ongoing relationship between the two.

The Cybersecurity Team

It may seem obvious but one of the most critical segments a CISO can have is their own team. Due to the criticality of this segment, it requires strong organization of the relationships to ensure that everyone on your team feels valued, engaged, and supported, in an ongoing fashion. You maintain these relationships already; let's enhance what we currently do by using our formalized SRM process to track our relationships with them to ensure no one is missed.

The relationships with your teammates will be some of the deepest and rewarding of all your connections. They drive the execution of your strategy and work tirelessly to advance the cybersecurity program. Because of this, I tend to keep track of more information on my teammates than any other connection. For example:

Birthdays: A small email or in-person birthday wish is something that will put a human aspect to what we do as leaders. It demonstrates that we are thinking of the person on their special day and goes a long way to personalizing our leadership styles.

General Information: Understanding basic information about our employees' lives outside of work helps to personalize our interactions with each of them. Information about the employee's spouse, significant other, children they have, the pets they enjoy, and perhaps hobbies and interests, all of these aid in customizing your conversations with them. In my career I've had the privilege of working alongside wonderful people, each one with unique characteristics. Extreme skier, guitarist for a band playing 1980s classic rock music, ultramarathon runner, ultimate fighter, and a threat intelligence specialist who lived on a goat farm – their backgrounds were fascinating to get to know and provided me with great conversation topics.

Work Specialties: Beyond our direct reports, understanding the key projects each one of our employees are supporting, as well as their area of expertise within cybersecurity and how it fits into the overall team is crucial. The best way to garner this information is through skip level meetings, a valuable tool in a CISO's SRM program. Consistently updating this SRM component will ensure you are up-to-date on recent training, certifications, or project successes by the employee.

As a CISO, you probably were already aware of the many different segments that require a strong relationship. Now that they are on paper, I'll bet you are surprised of exactly how many there actually are. Did I miss any? Of course I did! You will have other segments that may come to mind that are important to include in your strategy. You are well on your way. Now let's dive into some common attributes for each of your segments.

6
The Core Attributes of Segments

The entire purpose of segmenting your cybersecurity "consumers" is to provide specialization to how you deliver services to different groups with different needs. However, there are certain elements that are consistent from segment to segment that are worth mentioning. These common attributes should span all of your segments to some degree and should be included in your security relationship management (SRM) portfolios for each segment you've identified.

Critical Stakeholders

Within each of your segments there are those who impact the success of the segment so greatly, you'll want to ensure you highlight them so you can make the connection. Obviously, leaders will probably make up a large number of the critical stakeholders for the segment, but never lose sight of those stakeholders who may not be in a leadership role, but their knowledge and influence in a segment demands inclusion in your core attributes work. Instead of attempting to *track* all relationships with everyone in the segment (not *creating* relationships – you still want to expand your connections), I found success in identifying those critical stakeholders and tracking just those. You may find, as I did, that the influence of your connections often spreads the word of your cybersecurity initiatives throughout their organization.

DOI: 10.1201/9781003597483-6

For each of the target stakeholders, you'll want to collect and store information related to the person. I like to include preferred methods for contacting them, all contact information, background, hobbies, and many more details. You could add other notes related to the connection's cybersecurity fluency, influence in decision-making – whatever you'd like to include that would add value to the relationship you are nurturing.

Metrics and Measurements

Each of your segments benefit from measurements that are reciprocal. The information security organization should collect and measure, possibly through the company's ticketing system or other methods, the number of cybersecurity service requests for each segment. How often does a particular segment reach out with a security question or a particular need? This provides great insight for the CISO into how involved the segment is in your security initiatives and can serve as feeders into your demand management initiatives by anticipating future needs from the various segments.

Delivering metrics to the segment is equally valuable. For example, the CISO can provide to the leader of a particular segment things like number of clicks from that segment's employees in the cybersecurity phishing campaigns, number of critical vulnerabilities for the segment, patching performance percentage, number of vendor security reviews requested/completed, number of compliance issues, and a variety of other metrics – all focused on that segment. Because every segment may have different needs, customization of the metrics for a specific segment leader will provide real value for them.

Segment Contributions

Understanding how the segment contributes to the overall success of the company unlocks key details that can help focus your cybersecurity strategies. Business metrics, such as the percentage of revenue for the segment as it compares to overall corporate revenue, is one type of business metric that could provide significant value. Another important data point may be the largest customer or partner of the segment. Garnering data on these types of contributions the segment makes to corporate success aids in the security translation of which systems drive the most critical processes. Many of these data points can be identified through researching disclosure statements, or through interviews with key segment leaders (after all, they know their segment best). The CISO could supplement their research with information already in place; for example, the segment may currently use various platforms for their analytics and tapping into those could prove extremely helpful for the CISO.

Some of the contributions for segments may seem apparent to a CISO; for example, the sales department is obviously there to sell the company's products and services. However, a deep dive into how they sell the products, i.e., direct-to-consumer, channel sales, and strategic alliances, and the percentages for each, may not be so obvious and should be understood for maximum effectiveness of your security program.

For functional areas not necessarily in a profit and loss capacity, a narrative of how they enable the business to achieve effective results can be valuable as well. As an example, you may have a data analytics department that doesn't necessarily contribute directly to revenue, but their contributions and projects deliver analytics that assist in critical business decisions. Any measurements that help to clarify how the segment impacts the company through its contributions is an attribute the CISO must understand for each segment.

Segment Details

For each segment, a detailed understanding of the number of employees and segment location(s) is necessary. Information on the product lines, services, or functional capabilities each segment delivers will also add to a great overview of the segment.

In this portion of your SRM work, going above the minimum information suggested in this book is encouraged – the more information the better. In an upcoming chapter on the mechanics of SRM, we'll dive into how these data points can be leveraged effectively.

Feedback Loops

You will want to have an effective way (ideally several ways) to obtain, collect, process, and use feedback from each segment in your SRM program. Transforming this feedback into satisfaction metrics for the CISO paints a picture for what is working well, where the pain points are, and potentially introduce future opportunities, as they relate to the individual segments – and possibly, for the entire corporation. "Customer" surveys after services are delivered is one potential method. I lean heavily on interpersonal feedback; simply asking those within the segment how the security organization enables the work they do, or just as important, any impediments to their success as a result of our security initiatives. No matter the feedback, positive or constructive, the goal is to gather as much as you can (and from multiple sources) so you have an accurate view of the impacts from your security program.

It is important to be as specific as possible when asking your feedback questions so that the responses are directly applicable to future improvements of your program. Like SRM itself, feedback isn't an annual event – it requires constant collection and evaluation if it is to be successful.

Compliance Requirements

Identifying compliance-related items within each segment is another key attribute to gather. Various segments may have more direct impact to compliance success, for example, a part of the business that collects payment card or healthcare data. Another compliance component could be the effectiveness of the user access review program – identifying those segments who have challenges in achieving optimal compliance in this area. Understanding these impacts assists in targeting your compliance-related services where they can be maximally effective.

Cultural Considerations

There are cultural characteristics that may impact how we interact with stakeholders from varying regions across the world. I have been blessed in my life to have been exposed to several different cultures throughout the world. I've experienced first-hand the cultures in many countries in Europe, and even lived in Sicily for two years, as well as extended visits to Latin America, Asia, and the Middle East. I led the cybersecurity program for a Japanese company and was very excited about my first trip to Tokyo in 2013 to attend a global IT meeting. Prior to the trip, my company invested in a program that provided a class to U.S. employees on Japanese culture. We learned about the proper way to present and receive business cards, protocols for gift-giving, using mass transit, attending business meals, and other traditions. Even more important, we learned about the importance of preserving relationships and the heightened attention to the respect for others.

During one of the breakout sessions from the larger IT meeting, some colleagues from Tokyo and I were discussing security policy management and the upcoming implementation of our policies, procedures, standards, and guidelines to our company. In the United States and other areas, policies and procedures are considered mandatory while guidelines are offered as recommendations. After some initial confusion and further conversation, we learned that for this group of Japanese executives, guidelines were considered mandatory documents. It was a cultural difference that impacted the delivery of our security services, and we worked through that difference successfully.

Maybe you reside in Latin America and are interacting with a U.S. employee, or someone from France developing a working relationship with a colleague in Australia, there are cultural considerations that should be acknowledged whenever (and wherever) we move outside our own cultural norms. But international cultures are not the only ones to consider; within each company there may be cultural influences as well. Perhaps there is a business unit within your company that has a more deeply entrenched security culture than other business units in the same corporation? During my tenure at a global aerospace and defense company, there were nine operating sectors at the time, each with their own wonderful product and service specialties. Due to U.S. Government regulations, one of our operating segments developed more stringent security controls (both physical security and cybersecurity) that were beyond the already strong controls across the other sectors. While it may appear this would be a compliance attribute in your SRM program, my example is actually a cultural one. The advanced security controls were so deeply embedded in the culture of the sector; every employee understood the importance of the work they did and how it contributed to meeting the unique compliance requirements. There were two and three generation shipbuilders who were raised with this culture of security – it became instinctual.

Your SRM program, specifically within the core attributes of each segment, should contain a section on cultural considerations. Whether it is the traditional international type of relationship or a variation to the culture encased in your own company, identifying these considerations will maximize your program.

Documentation Retention

For each segment, I have found success in retaining the presentations I provide to (and receive from) the various stakeholders within it. It is also a logical place to house notes and work products related to that segment. In doing so, I have a record of past interactions to serve as a refresher mechanism for future discussions. I collect the documentation in the folder assigned specifically for that segment. A simple file organizational structure like this is easy to maintain and allows for quick retrieval of files.

Because so much of our work resides in electronic mail (e-mail), I leverage a similar folder structure as well in my email client – a folder for each segment. You can choose to go deeper with folders for specific stakeholders within a segment as well, if you choose. We are all very busy in our daily work and to make email sorting as easy and quick as possible, I leverage tags (or categories or labels) quite a bit in my email account. Each segment gets a specific tag or category which makes it simple to quickly assign a tag to the email for sorting at a later time if I am currently pulled away to address an issue.

The core attributes discussed here can help you identify and prioritize cybersecurity initiatives across varying levels of risk within your company. With this information, the CISO can perform focused risk management based on the unique needs and risks for each segment. And for the purposes of this book, greatly enhance the SRM principles that are so vital to deliver a high-quality cybersecurity program.

7
The ABC's of SRM

You have successfully built the foundations of your cybersecurity program, identified the services you and your team will provide, and created a branding strategy to promote them. In your segmentation work, you took the time to identify all of the segments you'll track, and gathered all of the core attribute data for each segment. Now let's work to put some action behind all of these endeavors.

The benefit of security relationship management (SRM) is clear: to provide a promotional strategy for your cybersecurity program by building strong relationships across all segments of your company. The more in-depth the relationship with each segment, the more heightened the awareness of your cybersecurity program and what it has to offer. Measuring and gauging how effective the relationship is with each segment starts by creating different levels of connection status. This is simply a measurement of how frequent your interactions are with a specific segment or person. How you differentiate the levels, as well as how many levels you have, is up to you; what follows is how I choose to distinguish between the tiers.

Connection Status

Simply put, the status of a connection, be it with an entire segment or individuals within each segment (ideally both), comes down to the number and quality of the interactions. The more the interactions, the higher the level of connection you will have with a stakeholder. Here are some suggested tiers.

DOI: 10.1201/9781003597483-7

Advocate

An *Advocate* is someone or some segment that you interact with frequently. Your discussions are of significant depth and relative to what is top of mind for the segment. Additionally, the relationship you nurtured with the segment is so effective that this connection is now an advocate for your security program. They seek you out for cybersecurity matters and may often support your initiatives in meetings that they attend – even when you are not present. These connections are the highest level of connection status you can have, and something to strive for in your other relationships.

Connection

The next level in your tiered program is simply called a *Connection*. This is someone or some segment that you have recurring conversations with, but perhaps they occur infrequently, or the depths of the interactions aren't as detailed as they could be. There is a bit more work to be done in developing the relationship to move it to the *Advocate* status.

Initial Interaction

Perhaps you've reached out to a segment stakeholder that you've not worked with previously and just had a very productive first meeting. This is the *Initial Interaction* level, and it is a promising indicator of work the two of you can engage in going forward. But if you are like me, other initiatives pop up during the week and soon you realize it has been two months since you've had that first discussion. By tracking your engagements, the risk of missed opportunities to connect is reduced and more engagement strategies can be created. The goal for the *Initial Interaction* level is to move the stakeholder/segment out of this level as quickly as possible.

Shadow

The lowest level of connection status is called *Shadow* and represents all of those persons/segments you've identified as someone you want to have targeted conversations with but haven't yet. Importantly, shadow doesn't represent *all* persons within a segment, that group may be far too great, no, it is just those people that you identified in your segmentation strategy in which you want to develop an ongoing relationship in order to enhance the promotion and delivery of your security services. The *Shadow* tier symbolizes untapped opportunities for enhancing the cybersecurity program, and work should be done regularly to continuously reduce the number of your identified stakeholders at the *Shadow* level.

Tracking Connections

Tracking projects, to-do's, shopping lists, and a surplus of other data is a normal part of our daily lives. Yet, we often don't devote adequate time in tracking our professional relationships to ensure we are providing the needed attention to them. For readers who can remember large-scale projects/experiences from your past, i.e., Y2K project, military service, team sports, and marching band, you most likely won't remember all of the detailed tasks associated with the experience, but you will remember the relationships you fostered during that time. Relationships matter – so why not track them the same way we do tasks and projects to ensure they are optimal?

SRM tracking consists of quickly notating various numerical values (points) relative to an interaction you have with your segment stakeholders, ideally right after it takes place. These points are based on the type of interaction, who it is with, length of the interaction, and other variables. The total number of points for each segment/stakeholder determines the connection status (*Advocate*, *Shadow*, etc.) for a segment/stakeholder outlined previously. It may sound confusing now, but in the next chapter we'll dive into three scenarios to show how it works in practice.

To capture points, one could use a simple spreadsheet, a notes application on a mobile device, or a project management platform, or track it with any other preferred application, with the goal of retaining a record of the point values for your interaction for later analysis. When I first started my SRM program, I attempted to use three different CRM technologies for my tracking but found they couldn't be customized enough for my cybersecurity efforts because they were more focused on sales and consistently reverted back to tracking leads and the closure of deals. Frustrated, I leveraged my own cloud-based solution (Figure 7.1) that worked well for the simplicity I wanted for quick entries that provided analytics for me to review at a later time (more on this in a moment). Regardless of the tool you use, the important thing is to capture the data somewhere.

It is worth mentioning that there is a distinction between tracking interactions with segments and tracking individual interactions. I start with tracking individual interactions because I found it to be easier to summarize the segment performance with the data obtained in tracking the conversations I have with stakeholders from that segment. Once I have the individual metrics, I can apply them to the segment in which they belong.

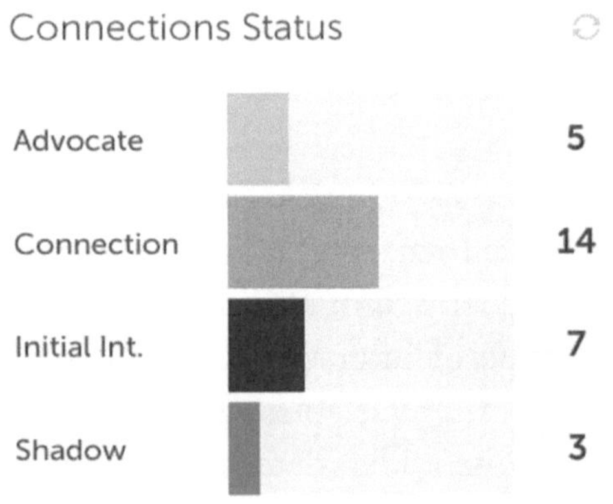

Figure 7.1 Connection Status.

The reason I do this is to avoid the mistake of having a strong relationship with one stakeholder in a segment but lack relationships across many of the other stakeholders within that same segment that I previously identified.

Collecting the tracking data is just the first step; without a regular review of the data, the tracking information provides little value. Each month, a review of the tracking data unlocks opportunities to understand:

- Key interactions accomplished during the past month;
- Identify prospects in the *Initial Interaction* level to set up more conversations for the upcoming month;
- Brainstorm steps needed to move a *Connection* level stakeholder to the *Advocate* level;
- Strategies you'll use to make an interaction with a *Shadow* level person.

Regular reviews of your SRM program turn data into action. This simple step, perhaps an hour of time during the month, is crucial to ensure consistent movement is made to nurture your professional relationships for maximum growth.

Connection Information

As you continually expand your growing network of relationships, it may become more difficult to juggle all the specific details for each of your connections. You'll want to personalize your interactions with each stakeholder, but with so many relationships, how can you keep up to date? The best way that I found is to detail the information about the person that will refresh my memory when the time comes to meet with them. I typically include:

- Name (including preferred name, i.e., Robert or Bob);
- Job role/Title;
- Company;
- Organization;

- Segment in which you placed them;
- Contact information: address, office number, phone numbers (mobile, office), email address;
- Their preferred method of contacting them;
- Assistant name and contact information (if applicable);
- Social and professional networks;
- Birth month and day (for birthday wishes);
- Notes section to highlight other information such as hobbies, family members, interests, and the like.

When talking with some of my peers in the cybersecurity industry, a few let me know that they like to track other information on their stakeholders. Some of them include notes on the level of influence the person has as compared to others at the company. Another idea would be to capture the connection's level of cyber fluency. The amount and type of data you include in your connection information is up to you – whatever nuggets of data you feel would assist you in your SRM endeavors. Knowing this information helps to personalize the time you spend together and sets the stage for a collaborative dialogue.

Top Initiatives

For each stakeholder, I include a list of initiatives that are top of mind for them so that I am sure to discuss each one during our time together, provide updates, or ask clarifying questions. These initiatives need not be solely cybersecurity related; the CISO should understand non-security initiatives that are important to the stakeholder as well. Knowing these helps to potentially unlock opportunities where there is a cybersecurity component that can be merged with what the stakeholder is trying to accomplish. By not knowing these initiatives, the CISO is limited in their ability to explore future opportunities, nor are they able to better understand drivers that may potentially conflict with cybersecurity efforts.

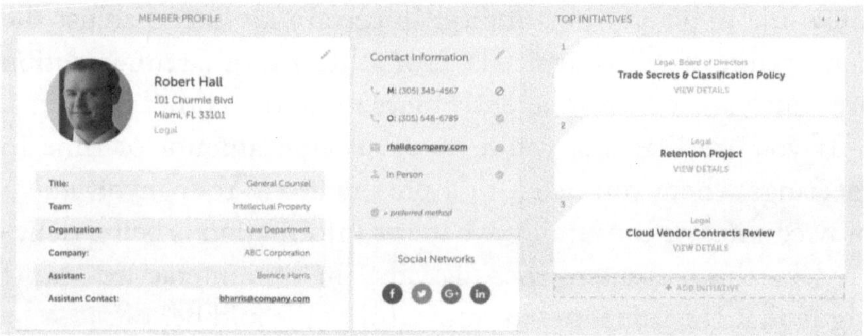

Figure 7.2 Connection Information.

When done with your collection of data for each stakeholder, it should be stored for quick and easy reference. An example of a connection information card is outlined in Figure 7.2.

Prior to meeting with any stakeholder, I review the information card to identify general interest information, i.e., how was the golf tournament they attended last week, as well as if any movement was made on an initiative since the last time we met. A quick review of this information before going into the discussion helps to personalize the conversation and makes the interaction more fluid.

We've established the importance of tracking your SRM performance to enhance the program's overall effectiveness, and what has been provided thus far in this book could put you on a great path to success. We could stop right now and have a good SRM program. But why stop now? Let's dive into some additional things that could elevate your program even more.

A Level Deeper

In a manner of speaking, you and I are communicating right now, albeit a one-way type of conversation. Nevertheless, the content of my message is being delivered to you through the pages of this book. Imagine if, instead of reading this, you and I were having a conversation about SRM over a cup of coffee. This would afford you the opportunity to ask me questions and would allow me to gauge your interest in certain areas and adjust the conversation to fit your needs. Not all forms of communication are equal; some hold inherently more value than other mediums.

If you are like me, you've spent a large amount of time in meetings where you listen to a presentation. Your involvement only consists of consuming whatever information is being delivered by the presenter. There is value in this interaction, but I argue that the value isn't as impactful as it could be.

As a reliable medium for decades, phones have allowed us to have meaningful conversations with business colleagues to discuss important topics. And with the advent of mobile phones, our interactions, and the convenience of them, exploded – making the phone a quick and easy way to share information or make a decision. There are limitations to phone usage however; we can't monitor the non-verbal cues from the participants. Additionally, there is a huge quality distinction in the value between a voice call and with a quick text message on the mobile device.

Since the start of the pandemic in 2020, our methods of communication were drastically altered, shifting us to video conferencing as a primary medium. For nearly two years, my close business interactions were confined to the screen of my 27-inch computer monitor. Yes, I could see the person I am talking with (if they have their web camera turned on), but video conferencing has its limits to effectiveness as well.

In-person collaborative meetings that assemble the participants in a room to discuss important matters allow for the exchange of ideas, challenging one another, as well as monitoring non-verbal cues from others. Even with the high-fidelity interaction found in an in-person meeting, certain variables make some in-person meetings more valuable than others. If too high, the number of meeting participants can easily dilute the effectiveness of the conversation. The smaller the number of participants, the more each attendee has an opportunity to engage.

Moreover, the setting of the meeting could impact the interaction. For example, a meeting conducted away from the office helps to strip away the distractions found in the normal office setting. Furthermore, interactions at a corporate social function where attendees can get to know one another in a casual setting could be positively affected as well. I've heard many times from security conference attendees that the content of the conference is good, but the real value is found in the discussions outside of the conference sessions over dinner or during a happy hour event.

The most effective way to build the relationship, convey important information, and listen to the concerns of the stakeholder is found in a one-on-one, in-person interaction that delivers the highest value over the aforementioned types of engagements. This is not to say that the rest of the engagements aren't encouraged; each one provides an opportunity to build the relationship and have varying degrees of significance.

But how does one track our total interactions and the variability considerations found in meeting types? The next chapter explores potential analytics to help quantify your SRM program.

8
SRM Analytics

As discussed in the previous chapter, not all interactions are equal. Yes, any interaction has the potential to advance your relationship, but there are inherent quality differences in them due to the structure and medium used to interact.

Since there are certain variables that make some interactions seemingly more valuable than others, it makes good sense then to include this variability in your security relationship management (SRM) points tracking and analytics work. After all, you are striving for maximum effectiveness of your program, and being as comprehensive as possible lends itself to a higher level of value from your SRM endeavors.

What follows are some tips I've used to differentiate between interactions and effectively track them so that I can heighten my relationships across all of my segments. The main goal for these analytics is to amplify the quality of your ongoing relationships.

I score each type of interaction I have with a connection in my SRM program using different point values. The higher the quality of the interaction, the higher the points I give to it. The types of interactions I track start with the email or text formats. This only includes messages sent and received between the connection and I and do not include emails/texts with a large distribution list of recipients.

The next variable I track is a phone call or conference call with three or more attendees, with an additional level of scoring for a one-on-one phone call with just the stakeholder and myself. The next seven levels are focused on

DOI: 10.1201/9781003597483-8

meetings and begin at large virtual meetings and ending with an in-person, one-on-one meeting with the connection – and several levels in between. Lastly, because the environment could potentially contribute to increased usefulness of the interaction, I give myself a multiplier of two to the standard points for any meeting conducted offsite, out of the normal office environment where distractions may detract from the conversation.

I found that by scoring my interactions this way, I am able to draw out more point values conducive to the quality of the medium used to conduct the interaction. This system motivates me to leverage more high-quality channels rather than firing off a single email to the stakeholder. I consistently look for ways to advance my quality scores, which, in turn, elevates my connections to higher tiers in my connection type model.

To illustrate the different point values I use relative to the type of engagement, see Figure 8.1.

Part of my routine is ensuring that I score an interaction immediately following it – while the details remain fresh in my mind. Waiting until sometime in the future to track the points introduces the risk that I may fall behind in my scoring, or that

Engagement Type	Description	Points
Email/Text	Originator of email/text or responding to an email/text between the CISO and the stakeholder only	1
Phone - Group	More than two participants on the call	2
Phone - 1:1	1:1 phone call with stakeholder	5
Meeting - Virtual - Group - Presentation	Attending a virtual meeting with a one-way delivery of content to participants without much interaction	3
Meeting - Virtual - Group - 10+	Attending a virtual meeting with ten or more participants in which there is interaction	4
Meeting - Virtual - Group - 3+	Attending a virtual meeting with three to nine participants in which there is interaction	5
Meeting - Virtual - 1:1	Attending a one-on-one virtual meeting	8
Meeting - In Person - Group - 10+	Attending an in-person meeting with ten or more participants in which there is interaction	6
Meeting - In Person - Group - 3+	Attending an in-person meeting with three to nine participants in which there is interaction	7
Meeting - In Person - 1:1	Attending an in-person one-on-one meeting	10
**In-Person Offsite Multiplier Bonus	For in-person meeting types held offsite, away from the corporate office, add a 2x multiplier to the standard points	2x

Figure 8.1 SRM Analytics.

the details of the interactions may be forgotten. It only takes a quick minute to record the points and some key takeaways from the interaction.

Every month, usually the last business day of the month, I tally the scores for each of my connections in my SRM program (or it is done automatically depending on the platform you use). My monthly goal is to have at least a single one-on-one meeting with the key stakeholders I've identified in my segmentation work, supplemented with a small collection of low point value interactions via email or phone calls. The points you set as monthly thresholds for each tier are dependent on the level of interaction you wish to have, so customization of the model is encouraged. In Figure 8.2, you'll see the values that I use.

Remember my example from an earlier chapter in which I shared that my knowledge of the Spanish language diminished over time because I wasn't using what I had learned? Each month that I wasn't immersed in the language, some of that knowledge eventually leaked away from my memory. The same holds true for SRM. With the passing of time, interactions that are not regularly cultivated lose their effectiveness. As a result, if during your monthly review you find that you missed an interaction with a stakeholder in that particular month, subtract ten points from that connection. This type of fluid give-and-take point valuation motivates me to maintain the relationships I've worked so

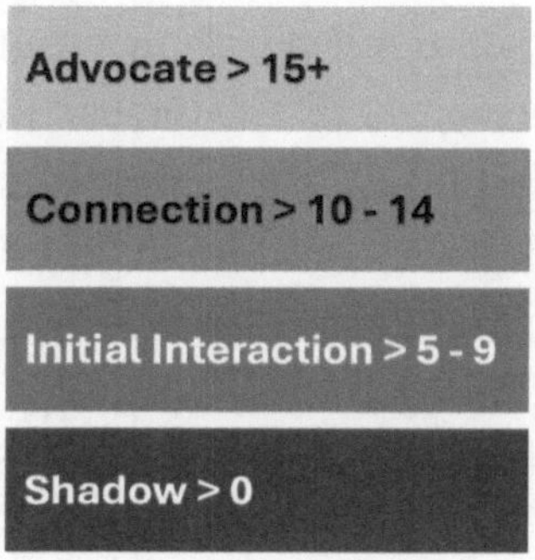

Figure 8.2 Connection Level Scoring.

hard to establish and doesn't allow me to get comfortable with my connection levels by just doing a running total of interactions to date.

Let's put these analytics to work by looking at a few scenarios.

Scenario #1

In my segmentation work, I identified the company's General Counsel as a target for a relationship that I want to establish and maintain – something that I have not done yet. I send him an email explaining that I have some ideas on how we can disclose our information security program in the 10-K filing for the end of the year. He responds with an email saying he would enjoy discussing this with me and I set up a one-on-one meeting using our video conferencing solution. The following week, we conduct the meeting and share some great ideas for our upcoming disclosure.

In this scenario, my connection with the General Counsel accumulated ten points for the month, (one point for my initial email, one point for his response email, and eight points for the one-on-one virtual meeting), moving him from the *Shadow* level to the *Connection* level.

Scenario #2

It is my goal to start an effective working relationship with the vice president (VP) of our supply chain operations. After an in-person strategy meeting (held in a local hotel ballroom) with fifteen participants, including the supply chain VP and myself, I approached her to ask if we could meet so I can learn more about her top initiatives. Three days later, we met for an hour, in person, in her office, and discussed the strategic goals for enhancing our company's distribution processes. I provided her with a general overview of the cybersecurity program, the meeting ended, and we agreed to meet again soon. The following day, she called me to say that she is looking to onboard a service provider for some technology enhancements to our fleet management platform and wonders if I could have my team look into the security practices of the vendor. I delivered the results of the review to her via email, in which she responded with a request for another meeting between the two of us to discuss another supply chain issue – which we do the following day for thirty minutes. The monthly tally for this connection is 39 points (6 points for the in-person meeting with ten or more people; a multiplier of 2 since it was held offsite for a total of 12; 10 points for the in-person one-on-one meeting; 5 points for the phone call; 1 point for my email; 1 point for her reply; and 10 more points for the second in-person meeting). This places this connection well into the *Advocate* level of connection and represents an executive who is very interested in cybersecurity.

Scenario #3

Since the communication department is such an important function for the company, as well as my cybersecurity initiatives, I initiated a connection with the senior VP of communications. In the previous month, we had two one-on-one meetings using our video conferencing platform to discuss an upcoming cybersecurity article that was going to be delivered to the entire employee population. She called later with a follow-up question about an acronym used in the article, which I explained. The points for the previous month totaled 21 (8 points for each one-on-one virtual meeting for a total of 16; and 5 points for the phone call). This placed her in the *Advocate* tier of connection for the month. However, in the current month, we had no interactions at all. Because of this, I subtract ten points which now moves her down a tier into the *Connection* level status with a new total of 11 points. I make a note for myself to reach out to her in the upcoming month so I can raise her connection back up to *Advocate* status.

Broader Themes

Moving from meeting to meeting throughout the week, we may often lose sight of the larger view of our entire group of relationships. Are we nurturing all of them? Are there certain connections that we spend a considerable amount of time with and why? Do gaps exist across the business where we don't interact as much as we should? We'll struggle to obtain an overall view of our relationships if we don't look at the broader analytics and the themes derived from them. This is why it is so important to use analytics to enhance your SRM program. As an example of an overall view, what follows in Figure 8.3 is how I differentiate the success (or opportunities) for my connections.

Now that you understand the work involved in tracking your interactions, it may be a good time to go back to your segmentation strategy to re-evaluate it. Did you bite off more than you can chew when you were identifying all of the potential segments? Did you miss important stakeholders within a segment? Perhaps you missed an entire segment? As mentioned earlier, SRM is a fluid process, so re-evaluation is welcomed, and necessary.

You probably have realized that the work involved in tracking your connections is quite significant; there is a considerable amount of effort involved to be sure. But consider the benefits of doing so and you'll agree that your ability to extend your cybersecurity program effectively across your enterprise is greatly heightened as a result of your work. What's more, SRM work can enhance your succession planning endeavors. Imagine a new CISO taking the helm of your program as you retire or move on to another opportunity, and you hand the incumbent your SRM program portfolio. It contains all of the connections and the valuable information you've collected about the business. They'll see which segments embrace the cybersecurity program and where opportunities exist to grow a segment's involvement. The new CISO will have a historical record of what information was shared with each stakeholder and when. There is incredible value in your SRM portfolio and can reduce months of work for

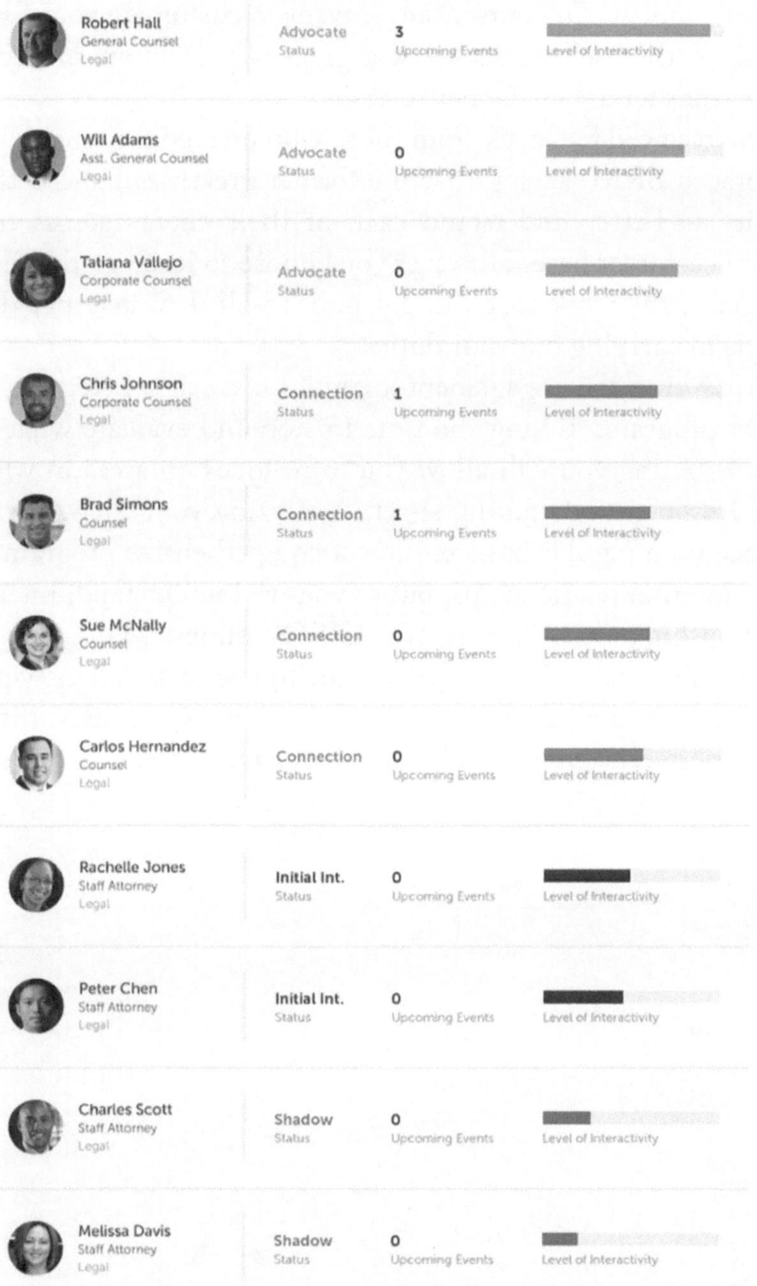

Figure 8.3 Overall Connection Levels.

the incumbent. Of course, they'll want to customize your SRM work for themselves, but what a great way to jump-start your successor's tenure as the new CISO.

I've received feedback from those who own consultancies that embraced SRM, saying that the format greatly enhanced their ability to better understand each of their clients across their entire customer base. However you choose to leverage the SRM program, the work that you put into it will definitely pay dividends in carrying out your duties.

Analyzing the engagement analytics is a critical step in your SRM program. Taking the time to stop and evaluate what the data is telling you will allow you to re-focus on areas in which you are potentially falling short. Analyzing your SRM performance on a regular basis ensures a comprehensive program for your internal relationships, but as you well understand, internal relationships are only part of a CISO's duties. Many external partnerships provide significant value to your cybersecurity program and in the next chapter, we'll dive into some SRM differences between internal and external engagements.

9
Moving Outside of Your Corporation

You have more than enough security relationship work to accomplish within your own company to keep you busy. But as a CISO, you understand that your internal relationships are only part of the total interactions you need to cultivate. There are security vendors, consultants, industry peers/groups, and a host of other professional relationships outside of your company that require nurturing if they are to be optimally effective.

For the most part, the security relationship management (SRM) elements that you have learned from this book thus far serve as a great foundation for tracking and preserving these important external relationships. Yet, there are some unique characteristics that should be considered when assessing relationships outside of your corporation.

Frequency

Generally speaking, I have found that interactions with many external relationships often do not have the same level of frequency as those we foster inside our companies. It may be the case that you meet with the external auditor, or perhaps a contracted security vendor, quarterly (or maybe less frequent than that). While the core elements of SRM data remain the same, you may find it more effective if you adjust your SRM analytics to specifically address this meeting infrequency among external segments/stakeholders. The reason is that SRM analytics derived from an infrequent meeting with an *internal* connection may show a less than effective relationship, while a quarterly

DOI: 10.1201/9781003597483-9

interaction with an *external* stakeholder may equate to an optimal relationship status. Customization based on frequency for these external relationships may be needed in your SRM program.

Scale

Within your company, the total number of potential connections is (for the most part) capped at the size of the company. Of course, employee count is fairly fluid based on turnover and increased fluctuations with hiring, but the pool of potential stakeholders has some fairly identified limits. With external relationships, the scale of potential relationships can (and often does) increase significantly year over year. Critically evaluating the selection of those external stakeholders that you *track* in your SRM program will help focus on those that are the most significant. As mentioned previously, scaling down only relates to the SRM measurement and analysis you perform, not limiting the number of casual external relationships you create.

Some examples of external relationships to track could include:

- Customer Advisory Boards in which you serve;
- Some of the security conference attendees you meet;
- Security think tanks;
- Your cybersecurity supplier partners;
- External auditor;
- Investors in security companies;
- Industry peers;
- And many more.

Data Elements

The types of data that you track within these external relationships include the basic components of the SRM program previously mentioned but may include other unique characteristics.

For example, when managing a relationship with one of your security vendors, you'll want to understand the contract you have with the vendor, what products you own from the vendor, total costs, contract renewal dates, and a host of other very specific data. Fortunately, there may be other systems you currently have that already provide this information. Regardless, I have found that understanding this information related to these connections, and in an easy-to-find format that I can track easily, has provided tremendous value.

Vendor Relationships

We rely on our security vendors to provide us with their technology and services to support our mission. Obviously, we should have strong relationships with those companies that serve as extensions to our cybersecurity team; yet there are so many other companies that we don't currently leverage who wish to do business with us. The sheer number of unsolicited connection requests from these external sources can be extremely difficult to manage.

In recent months I made a comment on a podcast about the number of vendors who reach out to me daily on various channels (email, social media connection requests, work and personal mobile phones, office phone, etc.). Some listeners, I assume they were not CISOs, who all too well understand the saturation of vendor requests, challenged my number, saying there are only so many security vendors, how can it be so many?

For each email from a hopeful vendor hitting my inbox, it is followed up by a social media connection request, a phone call to my work phone, and a phone call to my personal mobile phone. It isn't a one-to-one correlation between a vendor request and the number of cybersecurity vendors in existence. Occasionally (actually it happens a lot), I will receive an unsolicited email from someone telling me about the benefits of their platform and would like to meet with me to explore opportunities. However, the request comes from someone at a cybersecurity company in which I already do business (have for years) and have a named account representative. It isn't uncommon to receive the same exploratory request from multiple different email addresses from the same vendor.

Lastly, cybersecurity companies aren't the only vendors reaching out to us. There are huge numbers of requests to join regional conferences or industry meetups, as well as requests to participate in vendor product or industry studies. Many of these requests are sent three, four, or five times to "keep it at the top of my inbox" or "ensure that I didn't miss it". Some vendors will even summarize the number of times and methods they used in previous unanswered messages to me, i.e., "Hello Lee, I have reached out to you three times by email and left a phone message for you – I'd like to talk with you about…". I include this section on vendor relationships, something you are completely familiar with, not to lament on the number of requests I receive; instead, my goal is to highlight the problem of oversaturation in relationship building – something we will explore as part of the next chapter on addressing challenges.

We cannot run an effective cybersecurity program solely within the walls of our corporation. We require assistance from external partners, so why not track these important relationships? With some tweaking of your SRM process for these external connections, you'll be in a better position to maintain productive relationships with this important group.

10

Addressing Challenges

The effectiveness of our role as CISO may (in some cases) be limited because of our inability to have frequent interactions with the top corporate executives of the company – at least at the same level in which those executives interact amongst themselves. Leveraging an effective security relationship management (SRM) program allows us to insert ourselves into more regular conversations by our own initiative – rather than waiting for others to invite us to interact. What happens though if, despite our best efforts, our requests to engage with a stakeholder go unnoticed?

Similar to marketing endeavors, our work with SRM doesn't promise consistently successful results. An interaction takes two people to make it work so there may be times that, even with well-organized planning, our target stakeholders may not engage with us as much as we would like. Moreover, the recipient may have pre-conceived mindsets (positive or negative) about certain things, like cybersecurity, that could impact their willingness to engage with you more frequently. What follows are some tips that I've tried in my career where I've had some success.

Managing Engagement Challenges

1. *Alignment*: As mentioned previously, cybersecurity must be aligned with the goals and desires of your stakeholders if it is to be successful. In marketing terms, I believe this addresses the "wants" aspect of consumers. By making security something that is desired by your stakeholder,

DOI: 10.1201/9781003597483-10

they are more likely to engage to learn more. This is accomplished by stepping outside of what you (as the CISO) are trying to achieve and by presenting your program in such a way as to support the stakeholder in accomplishing their goals. We often call it "enabling the business" and it creates a direct link between our cybersecurity goals and the goals of the business entity in which you are interacting. Make your cyber message personal to them to create a "want".

2. *Accentuate the Necessity of Security*: Aside from the "wants" aspect of consumers, there are certain "needs" that we all have – safe and secure computing is one of those needs. Without it, our companies are less likely to effectively engage in business initiatives with reduced levels of systemic risk. In the back of their minds, our stakeholders are aware of this importance, but it stands to reason that reminding them of this need for the overall good of the corporation should be a part of your strategy in breaking through any barriers to an effective relationship. I call it a "sophisticated level" of FUD (fear, uncertainty, and doubt). Similar to "wants", the more you connect the "needs" to the stakeholder, the better. Which risks are more likely for their area of the business, which systems and processes are vulnerable, and so on – examples such as these help to make the necessity of security relevant to them.
3. *Leverage Your Network*: There are times in which we all could use a bit of help. By leveraging mutual connections between yourself and the reluctant stakeholder, it may help to open up conversations. When another person publicly validates the benefits of their relationship with you, the reluctant connection may be intrigued and become more likely to engage with you to learn more about your message.
4. *Be Present*: At one of the companies I worked for in my career, I was promoted to an enterprise position

responsible for creating a holistic information security program out of nine disparate corporate programs. The role was to be located in the Washington D.C. area and my family and I were ready to embark on a new life, in a new city. Prior to beginning the relocation work from Mississippi to Washington, I attended a Washington Nationals vs. Philadelphia Phillies baseball game with my boss one summer evening. We chatted about my new role, and he asked, "So, are you ready for the move to D.C.?" Without any preconceived thoughts on the matter, I spontaneously answered that I was indeed ready but wondered out loud if D.C. was the best place for this role. At the time, the CIO of the company, as well as a large portion of the IT organization, was located in the Dallas, Texas office. The lion's share of the security team was in D.C., my boss lived in Pennsylvania, so I asked him, "Wouldn't it be better if you had this role in Dallas, close to the CIO and the company's IT organization so you have someone to represent cybersecurity locally?" He asked if I had family and friends in Dallas and I told him no, I didn't know a soul in Dallas. He liked the idea and two months later, my wife and I relocated our family to Texas.

The move proved to be an effective one insofar as I was present in the Dallas office and could interact with senior IT leadership on an informal basis several times per week. One strategy I used is what I referred to as "walking the halls". About three times each day, usually during breaks between meetings, I would take two minutes to walk through the executive wing to stretch my legs. Sometimes I would stop to chat with one of the executive assistants or an executive IT support employee, or perhaps pick up something from the shared printer. While it didn't occur each time I walked the hallway, on more than a few occasions an executive would pass by, and we would say hello and chat briefly. Every once in a while, an executive would see me in the hall and ask that I step into their office to ask a security-related question. My presence helped to increase my interactions by removing the formality of scheduling meetings and instead, leveraging quick, opportunistic interactions to help boost the relationship. These informal chats then led to future scheduled meetings with the executives.

Ten years after my time of walking the halls in the executive wing, the global pandemic changed the way we all interact. During this very difficult time, being in the office was no longer feasible. Today, a few years after the peak of the pandemic, many find themselves in a hybrid environment where daily, face-to-face, office interactions are still not as commonplace as the pre-pandemic era. Regardless, I still find value when I am in the office (in a hybrid environment) to roam the hallways to entice informal relationship building.

Oversaturation and Balancing the Message

Most corporate communication teams are very cognizant of message overload to their employees. Often, they will schedule the delivery of the messages to ensure there isn't an oversaturation of communication to the employees. At one company I visited, there was a full-time communications employee who had the sole responsibility of measuring the number of communications being sent to employees at the enterprise level, how often specific topical areas were covered in messages, as well as which specific segments of the corporation the messages were delivered. They study the retention effectiveness of the messages received and balance the delivery timing to ensure maximum retention.

With an eye to prevent oversaturation, your SRM program should be similarly cognizant of the number of messages being sent to your connections. It is true that we want to engage frequently to nurture the relationship, but not at the expense of appearing bothersome. By tracking your interactions using the methods outlined previously in this book (customized to your program), you will be able to more effectively control the optimal frequency of your messages for each connection.

Unconventional Measures

After the effective completion of a task or project, or perhaps a very productive meeting, it is always nice to receive an email, phone call, or in-person "thank you" from someone. I have found that going a step beyond a quick email can help to strengthen relationships and increase the willingness of a connection to engage more with you in the future.

Consider keeping a box of thank you cards in your desk drawer so that from time to time you can write a handwritten note to a connection. Taking the time to hand-write a message goes a long way in nurturing relationships. I like to keep a stack of gift cards from the local coffee shop so that I can either give one to a connection who is greatly supporting cybersecurity, or perhaps take them out for a coffee (the latter provides yet another opportunity to connect with that stakeholder – make sure you track this offsite interaction in your points tracking!).

Lastly, everyone enjoys a free meal; buying lunch for a connection from time to time helps to boost the relationship you have with them. The limitations of a hybrid work environment can somewhat impede our ability to have a coffee or lunch together. I suggest a scheduled meet-up to accomplish these so that the "in the office" schedules align between the two of you.

Geographical Challenges

Nearly every corporation conducts business, or relies on operational support, outside of their region. You may encounter challenges when nurturing a relationship with connections that are spread out all over the world and across multiple time zones. Remote work requires a hyper-sensitivity to those not in our location. All too often I have seen a meeting conducted in a corporate headquarters conference room that, besides the in-person attendees, included a few remote participants on the conference call line. For the bulk of the meeting, the employees around the table share ideas and in the last minute of the meeting, will ask, "does anyone on the line have anything they want to add?". We should include all participants in the conversation throughout the meeting, regardless of where they are in the world. I believe corporations have improved their engagement practices over the years, but geographical challenges remain a large barrier to overcome.

When decreased travel budgets occur, alternative methods for your in-person interactions help to maintain quality relationships, despite the distance between you and the connection. One simple method is to periodically adjust your schedule to allow for video conferencing for those on the other side of the world. With some effective scheduling, several interactions can be aligned in one day or evening over the course of a couple hours.

Previously I mentioned that email communications can be effective, but do not possess the same quality as an in-person interaction. However, during tight budget constraints, email may be one of the few options we have. Frequent, brief emails to say hello and to convey information is better than no communication at all.

Of course, obtaining sufficient budget to travel for in-person interactions allows for the most high-fidelity type of engagement. It is a wonderful way to experience the work environment of your stakeholder, obtain first-hand knowledge of business practices, and refresh interpersonal bonds.

Company Size

An argument can be made for not needing an SRM program if the CISO works for a very small company – even more so if that company is focused solely on domestic or regional business with a small number of employees. Why would I need to track my engagements if there aren't many employees that I interact with; we have very solid relationships because we are so small? Let's explore some ways SRM can help enhance their cybersecurity programs – despite the small size of the organization.

As previously highlighted, a well-executed SRM program focuses your interactions on what is top of mind for each stakeholder. As you nurture, and track, these relationships, the CISO obtains a more holistic view of business strategies – across all disciplines (functional and lines of business). This valuable insight benefits CISOs for companies of all sizes, and while one may have a more frequent and clearer view of the business in a small company, formalizing the interactions provides significant benefits.

If a smaller company has a hyper focus on keeping operational costs low and to maximize gross profit margins due (in part) to its size and market share, it may be the case that the security budget is limited, and controls rely heavily on its employees to deliver an effective cybersecurity program. If this is the case, the people aspect of the security program becomes that much more important, and an effective SRM program helps to enhance and nurture those relationships.

Even though the internal connections may be limited to a small number due to corporate size, the CISO has hundreds of external relationships to manage with partners, vendors, industry colleagues, boards, and so on. SRM helps to manage those relationships and would benefit CISOs tremendously, even those in a small corporation.

Lastly, building and maintaining an SRM program can provide valuable training on how to increase security alignment to the business. If or when a CISO moves into another CISO role for a larger, global corporation, they will have already developed the experience of SRM which will serve them well as they build the foundations of their new program.

You will have your fair share of challenges as you pursue great relationships, but with some innovative measures to address those challenges, your SRM program can remain effective.

11

The Future of SRM

Implementing a security relationship management (SRM) program to track your interactions across all corners of your business, as well as external relationships outside your corporate walls, involves a tremendous amount of upfront work. At its beginning stages, you'll find there is a lot of effort put into segmenting, creating the relationship tiers, developing your SRM analytics, and filling the program with meaningful content related to each of your stakeholders. Once the foundations are in place and tracking your interactions becomes commonplace in your daily routine, the amount and types of upfront work involved are replaced with sustaining an operational SRM mode.

It's not unlike taking the reins of a CISO role for a company. The initial stages involve assessment work, building a strategy, communicating the strategy, procuring technology, hiring staff, and creating an organizational structure in which to operate. The early months are filled with tremendous foundational work, ultimately giving way to increasingly steady-state operations. Of course, shifts in strategies due to evolving risks often take place, as well as new corporate endeavors (perhaps an acquisition or maybe a large technological shift), requiring a constant re-evaluation of your program. An effective SRM program is very much the same thing. Just because the initial foundational work is completed doesn't mean the bulk of our SRM effort is complete.

Ultimately, an SRM program is a tool that, once set up, provides you with actionable information on the performance of your relationships, as well as opportunities for areas in which

DOI: 10.1201/9781003597483-11

work is required. Our goal as CISOs is to align with the business and to enable it, and the process of SRM unlocks opportunities to accomplish these things. As we deepen our professional relationships, we obtain firsthand knowledge of business goals and the challenges that hinder our companies from achieving them. Following the SRM model and documenting important notes, we are now as close to the "customer" as we can get; our ability to understand the business more holistically is heightened, and our contributions to corporate success will follow.

The role of CISO is an extremely interesting one insofar as our involvement with projects across the entire breadth of our companies gives us the privilege of having a deep enterprise view. Moreover, the importance of cybersecurity has elevated our interactions with senior leadership and the board of directors. I believe the combination of these attributes positions our role as one that could extend beyond our current span of control, to contributing more to overall corporate decisions as a peer executive to the leadership team. While some industry analysts are calling for the reduction of scope or elimination of the CISO role, I see a bright future for our position. To capitalize on this opportunity, building solid relationships is the cornerstone to success. And SRM, I believe, is the key driver to do just that.

Thank you for your time in reading this book to explore how leveraging marketing techniques and SRM may enhance your cybersecurity initiatives. My hope is that you found some tips that could provide value to the important work you do each day by securing some great relationships. Similarly, I would value *your* feedback on the concept of SRM, successes you have enjoyed, challenges you may find, and how you are deploying it in your own way. You can reach me at www.novelsecurity.com with questions or comments. Let's have a conversation and build a relationship.

Appendix: SRM Toolkit

The following toolkit is offered as a reference for: 1) developing a few discussion topics with each of your segments and 2) potential channels to build and maximize industry-specific relationships.

SRM Discussion Topics

This is not a complete list; instead, it offers a few topical tips to help initiate thoughts on areas you may want to include in your conversations with stakeholders. You know your organization best, so customization of this portion of the toolkit is required if it is going to be effective.

Board of Directors	
Board of Directors	Convey management's use of due care in identifying, prioritizing, and addressing cyber risks to the organization. Create each presentation to the board such that they build off of one another. Each presentation is just one chapter in a larger book. Highlight material security events with commentary on what occurred and strategic plans (and completed milestones) to reduce the risk of future, similar events. Discuss the company's layered security model, each control, and the diversity in technology vendors, as well as how they all work together as a holistic security program. Track progress on the execution of the information security program, new and shifting initiatives as a result of emerging threats (external as well as internal business strategic changes). Answer questions related to industry breaches and your company's susceptibility to a similar attack/response (particularly with victim companies within your industry). Current and emerging security news, i.e., regulatory. Consider a special topic for each of your briefings, i.e., multi-factor authentication posture, artificial intelligence, and security of your financial systems (Hot tip: I recommend putting the special topic at the end of the briefing to keep the agenda on track). An overview of your tabletop exercises, things that went well, and opportunities to enhance the incident response program. Leave time to answer board/committee questions. Leave time to receive the board's/committee's oversight remarks. Frame your interactions with what the management team has done and its progress to open items. Focus on providing information that allows for board oversight rather than going to the board for solving issues.
CEO and Executive Team	
CEO and Executive Team	Communicate cyber risks for the company and the current controls in place to address those risks, identifying any gaps. Discuss the cybersecurity strategic plan that is balanced with business needs to address management's cyber risk tolerance. Communicate and garner insights from the various members of the executive team, i.e., CHRO, CFO, CIO, and General Counsel, on the specific security items related to their specific areas. Ongoing discussions on emerging risks that impact the entire corporation. Security budget/funding.

	Human Resources
Human Resources	Qualities and skills (and your gaps to those) that are essential for an effective information security program. Job role standardization of cybersecurity roles for effective compensatory benchmarking. Attracting top cyber talent strategies. What makes for an effective cyber internship program? Geographic locations of cybersecurity talent to optimally deliver services.
	Legal
Legal	Incident response to include notification, determining materiality, communication to board of material incidents, and other related processes. SEC (U.S. Securities and Exchange Commission) disclosure work. Industry work by the CISO to ensure no conflict of interest. Privacy-related matters (if no dedicated privacy organization). Fraud-related matters (if no dedicated fraud organization). Record retention. Data loss prevention, data classification. Response under privilege processes. Liaison with outside counsel.
	Corporation
Corporation	InfoSec Compliance Training. Centralization of security program (or detailed RACI for distributed InfoSec organizations). Enterprise security metrics. Cyber insurance work.
	Cybersecurity Governance Committee
Cybersecurity Governance Committee	Set the direction for the information security program. Dictate security policy requirements. Allocate resources to the information security program. Arbitrate conflicting business priorities requiring cybersecurity support. Develop and set the governance committee agenda. Prepare for upcoming board or board committee update.
	Information Security Champions
Information Security Champions	Creation and ongoing maintenance of the information security champions program, i.e., recruitment, training, and feedback loops. Security message distribution to the enterprise.

Other Committees	
Other Committees	*Disclosure*: SEC regulation, 10K, 8K disclosures. *Disclosure*: Determine materiality of cyber incidents. *Crisis Management*: Liaison with Federal, State, and Local law enforcement. *Business Continuity/Disaster Recovery*: Annual testing, continuity plans, business impact analysis, prioritization of recovering systems, etc.
Line of Business (Business Unit, Sector, Division, Segment)	
Line of Business	Receive information on unit's strategies, successes, challenges, and other unit-specific information. Provide unit-specific information security data/metrics. Unit's capabilities and how it contributes to the overall success of the enterprise. Unit specifics such as locations, key stakeholders, leaders, and customers.
Manufacturing	
Manufacturing	Key suppliers and methods for system maintenance. Unique equipment used, i.e., robotics, specialized engineering systems, and laboratory equipment. Network segmentation with core corporate network. Inventory methodologies and inventory levels. Number of units produced per day, week, and month, on average. Shift turnover processes. Maintenance windows and other line downtime.
Distribution and Shipping	
Distribution and Shipping	Logistic information on shipping of products. Technology partners for key technologies such as transportation tracking systems. Shipping vendors, transportation fleet details, wholesalers. Supply chain risks, including non-cybersecurity risks.
International	
International	Gain familiarity with scope of regional and country leaders. International cybersecurity regulations, particularly how they impact the region and the company overall. Key international partner, supplier, and government relationships for the country or region. Achieving balance between providing local support and enterprise security service delivery for cybersecurity resources in country. Identify and discuss cultural variables that may impact the delivery of security services. Explore any varying risk tolerances associated with the region or country.

	Finance
Finance	Security budget: Discuss detailed information regarding capital vs. expense; timing of investments; and other specifics. Possibility of setting up labor cost codes for tracking labor hours dedicated to incident response, prior to the need for response. Financial deep dives into corporate card purchases of software and cloud services outside of the IT support. Deeper understanding of the corporation's finances, challenges, strategies, etc.
	Sales
Sales	Explore major accounts, partners, and biggest revenue generators among the lines of business. Top SKUs (stock keeping units) from a sales perspective. Types of sales strategies and which are the most successful: direct to consumer, named accounts, strategic alliances, etc.
	Information Technology
Information Technology	Strategies for enterprise architecture. Strategies and performance of vulnerability management, patching. Strategies and performance of cloud computing, with emphasis on security. Strategies for network to include firewalls, wireless, etc. Details of core image for corporate systems, potentially multiple types of systems depending on employee job role, etc. OT and IoT strategies, securing those infrastructures. Compare with interactions from your manufacturing stakeholders. Email security posture and strategies.
	Digital and eCommerce
Digital and eCommerce	Impacts to the team related to PCI compliance. Deep dive into the Software Development Lifecycle and secure by design programs. Identify conflicts, if any, with speed to market of digital products as a result of security testing. Security awareness for developers. Explore omnichannel capabilities and the security of those channels.
	Communications and Investor Relations
Communications and Investor Relations	Scheduling security messaging to the enterprise, managing the message, timing of communications. Deep dive into the communication strategies for incident response: press releases, holding statements, notification (both external to the public, partners and internal to the employees).

Internal Audit	
Internal Audit	Ongoing tracking of audit items related to information security. Building the quarterly Enterprise Risk Management report for leadership. Suggesting areas of risk that could be audited to validate the risk posture of the company to a specific risk area. Partner with external audit on significant findings. Partner with external audit on quarterly information security reporting of material incidents.
Top Cybersecurity Vendors	
Top Cybersecurity Vendors	Ongoing discussion with the Managed Security Service Provider on the state of the partnership, areas of opportunity, etc. Contract discussions and budget negotiations. Establish a feedback loop on product/service effectiveness. Compare and contrast security team's strategies with the vendor's product roadmap to identify synergies and conflicts.
Cybersecurity Team	
Cybersecurity Team	Discuss employee activities and interests to drive more engaging conversation. Understand key priorities in their weekly security work. Discuss if they have the tools and resources needed to be effective in their role. Explore the employee's expertise in various security domains.

Industry-Specific Relationship Building

What follows are some channels for building and maintaining industry relationships – not a complete list, but a start to serve as a thought-provoking guide for you to establish your own strategy.

Cybersecurity Event Organizers
There has been a tremendous increase in companies that organize and deliver events to assemble cybersecurity leaders and practitioners to discuss various security topics. These can take the form of a dinner, ballgame, wine/bourbon tasting, or other unique events. The company may often leverage sponsorships from cybersecurity vendors to help absorb some of the costs associated with putting on the event.

Insights	These are a great way to build connections with industry peers, and more importantly, peers in your geographic region. The best connections are the ones in which you continue the conversation one-on-one in the months after the event. I like to focus on the content of the meeting by evaluating the agenda prior to the event to ensure it is something I have an interest in and can learn best practices from my peers on those specific topics. Some organizers will require that you select a specific number of vendors from a list of companies in order to have a meeting on each one's products and services – in between general sessions. If your goal is to obtain more information on cybersecurity products and services, this could be of value. However, if you are looking for more industry insights, the vendor portion may not be appealing to you. Always look for those organizations that follow Chatham House Rules at the events to ensure your input is kept between the participants. While not always an option, selecting a vendor-free event allows for a 100% focus on the topics, without consuming time on vendor discussions.

Award Programs

Another popular industry engagement channel is with companies that combine security content with awards for cybersecurity professionals. The awards are presented to CISOs, Cybersecurity Teams, Innovative Security Projects, and other security-related elements, usually sorted by geographic region. During the event days, there are sessions on various information security topics and the awards ceremonies are held in the evenings. Most of these programs have a large vendor presence as sponsors; I've seen small showcase expos during these events for attendees to learn more about each vendor's products and services.

Insights	These provide a wonderful opportunity to meet with many of your peers to discuss cybersecurity issues and get to know one another. Because these are often regional, the interactions with peers from the same location increase the opportunity to interact more in the future, outside of formal industry events. Many of these events highlight all of the award finalists and describe details of the great things they are doing at their companies – a wonderful way to gain some tips for your own program. These companies often put on social events at large security conferences at venues close to the conference event. These are great opportunities to re-connect with acquaintances and make new ones.

Information Security Conferences

Conferences are a popular venue for sharing industry-related information and meeting connections. They range from annual mega conferences to small quarterly regional conferences.

Insights	Many conferences charge a fee to attend and have various levels: full access, expo only, etc. When budgets are tight, it is appropriate to understand which conferences will provide the most value for you and your team. Don't rely solely on large security conferences as your only path into industry engagement. They are valuable, but it may be unrealistic for security professionals to attend many paid conferences in a given year. Consider speaking at conferences or participate on program committees. Most conferences will provide a free-all access conference pass to their speakers. It is also a great way to promote your own personal brand, as well as a way to meet other industry peers. If many of your teammates are attending the same conference, it is fun to host a dinner for them to socialize outside of the office, and the conference. It's also a good idea to invite some of your industry colleagues to the dinner so your teammates can meet other security professionals. Definitely attend the conference sessions because there is a lot of great content at the conference. But do consider spending significant time outside the conference by attending dinners and happy hours. These engagements are a valuable way to meet your peers (Hot tip: plan your evening calendar for the week well in advance of the conference). Most security vendors do an annual user conference to discuss their solutions and bring together customers. To bypass event costs, consider leveraging contract renewal time to potentially negotiate a certain number of free passes to the vendor's conference for your team as part of the renewal terms for the life of the contract.

Information Sharing and Analysis Organizations

These industry-specific organizations deliver relevant security information for its members and provide a community for security professionals within a specific industry.

Insights	As a member, ensure you are getting the most of the membership by engaging regularly and leveraging all of the sharing capabilities, events, and peer exchanges. Encourage your team to do the same.

Academia

Many universities have stood up cybersecurity programs, think tanks, and other offerings for their students, as well as the community. It is a great way to participate at the junction of industry and higher education.

Insights	Search for some Adjunct Professor opportunities at your local college or university. I have found many of them are enthusiastic to include industry experts as instructors in order to bring real-world relevancy to textbook knowledge. Spend time with the professors and other industry participants to share information. Think about offering your time in university mentorship programs or advisory boards.

Specialty Cybersecurity Communities	
There are some incredible industry communities for cybersecurity professionals who share specific attributes. There are cyber groups for veterans, women, and so many others that bring together professionals to discuss cybersecurity, but also how they can advance our industry.	
Insights	Think outside of your own participation. For example, if you are a male CISO, provide a corporate membership to a women's cybersecurity organization for the ladies in your organization. The same holds true for other specialty cyber communities; even if you are not part of that demographic, you can help by offering memberships for your team. In doing so, you'll be advancing our industry by supporting diversity – which breeds innovative thought.

1:1 Industry Peer Connections	
A one-to-one connection with an industry peer is one of the most valuable ways to gain industry insights. The connections are impactful because they are an informal discussion, void of distractions or vendor pitches.	
Insights	Don't rely only on social media or professional networking sites to nurture the relationship. Meeting over coffee, dinners, taking in a ballgame, or any other in-person engagement add tremendous substance to the relationship.

Index

Pages in *italics* refer to figures and pages in **bold** refer to tables.